NUTRITION AND DIET RESEARCH PROGRESS

HEALTHY FOOD

PERSPECTIVES, AVAILABILITY AND CONSUMPTION

NUTRITION AND DIET RESEARCH PROGRESS

Additional books and e-books in this series can be found
on Nova's website under the Series tab.

NUTRITION AND DIET RESEARCH PROGRESS

HEALTHY FOOD

PERSPECTIVES, AVAILABILITY AND CONSUMPTION

ANTHONY E. WALTON
EDITOR

Copyright © 2020 by Nova Science Publishers, Inc.

All rights reserved. No part of this book may be reproduced, stored in a retrieval system or transmitted in any form or by any means: electronic, electrostatic, magnetic, tape, mechanical photocopying, recording or otherwise without the written permission of the Publisher.

We have partnered with Copyright Clearance Center to make it easy for you to obtain permissions to reuse content from this publication. Simply navigate to this publication's page on Nova's website and locate the "Get Permission" button below the title description. This button is linked directly to the title's permission page on copyright.com. Alternatively, you can visit copyright.com and search by title, ISBN, or ISSN.

For further questions about using the service on copyright.com, please contact:
Copyright Clearance Center
Phone: +1-(978) 750-8400   Fax: +1-(978) 750-4470   E-mail: info@copyright.com.

NOTICE TO THE READER

The Publisher has taken reasonable care in the preparation of this book, but makes no expressed or implied warranty of any kind and assumes no responsibility for any errors or omissions. No liability is assumed for incidental or consequential damages in connection with or arising out of information contained in this book. The Publisher shall not be liable for any special, consequential, or exemplary damages resulting, in whole or in part, from the readers' use of, or reliance upon, this material. Any parts of this book based on government reports are so indicated and copyright is claimed for those parts to the extent applicable to compilations of such works.

Independent verification should be sought for any data, advice or recommendations contained in this book. In addition, no responsibility is assumed by the Publisher for any injury and/or damage to persons or property arising from any methods, products, instructions, ideas or otherwise contained in this publication.

This publication is designed to provide accurate and authoritative information with regard to the subject matter covered herein. It is sold with the clear understanding that the Publisher is not engaged in rendering legal or any other professional services. If legal or any other expert assistance is required, the services of a competent person should be sought. FROM A DECLARATION OF PARTICIPANTS JOINTLY ADOPTED BY A COMMITTEE OF THE AMERICAN BAR ASSOCIATION AND A COMMITTEE OF PUBLISHERS.

Additional color graphics may be available in the e-book version of this book.

Library of Congress Cataloging-in-Publication Data

ISBN: 978-1-53617-599-8

Published by Nova Science Publishers, Inc. † New York

Contents

Preface		vii
Chapter 1	What We Know So far About Orthorexia Nervosa: A Review *Márton Kiss-Leizer and Adrien Rigó*	1
Chapter 2	How Healthy Foods and Early Feeding Practices Can Be Effective in Primordial Prevention of Non-Communicable Diseases *Motahar Heidari-Beni and Roya Kelishadi*	21
Chapter 3	Functional Ingredients and Food Neophobia towards Healthy Meat Products Enriched with Agroindustrial Coproducts Flours *Nallely Saucedo-Briviesca, Alfonso Totosaus and M. Lourdes Pérez-Chabela*	39
Chapter 4	Candelilla Wax as Oil Restructuring Agent as Fat Replacer to Formulate Healthy Cooked Sausages *Alfonso Totosaus, Aislinn N. Botello-Pérez, D. Aurora Hernández-Domínguez, Octavio Toledo and B. Mariel Ferrer-González*	61
Bibliography		77

Related Nova Publications 159

Index 167

PREFACE

Healthy Food: Perspectives, Availability and Consumption first explores the pathological form of healthy eating, orthorexia nervosa. Although orthorexia nervosa cannot be found in the *Diagnostic and Statistical Manual of the American Psychiatric Association*, important findings suggest that orthorexia nervosa should receive wider scientific and public attention.

Additionally, the current literature regarding the effects of healthy foods and early feeding practices in childhood is explored in the context of the primordial prevention of non-communicable diseases and their risk factors.

This compilation also examines carrot bagasse flour and banana peel flour for their potential to be employed as functional ingredients to improve the texture, color, and flavor of raw meat products, as chorizo, or cooked meat products, as sausages.

Candelilla wax oleogel is investigated for its potential to replace pork back fat lard in cooked sausages, specifically focusing on its textural profile, moisture, color, and sensory acceptance. Results indicate that candelilla wax oleogel can be employed as a fat replacement, improving the health profile of certain meat products.

Chapter 1 - There is an emerging focus on healthy eating in Western culture. Although healthy eating is undoubtedly a core element of healthy life, this behavior can also become excessive and harmful. Being obsessed with healthy eating may lead to eating disorders and, paradoxically, the aspiration to be healthy becomes unhealthy. This phenomenon well represents how eating disorders are influenced by the actual social-cultural environment; recently the ideal of being healthy has gained strength besides the ideal of being thin, which may have contributed to the emergence of the obsessive type of healthy eating (Varga, Dukay-Szabo and Túry, 2013). The pathological way of healthy eating – orthorexia nervosa (ON) - has been described by Bratman (1997) and due to the relevance and actuality of the topic, the number of scientific publications are increasing in this area. Although ON cannot be found in the Diagnostic and Statistical Manual of the American Psychiatric Association (DSM-5) and it is still not fully understood, important findings have suggested that ON should get a wider scientific and public attention. The aim of this paper is to provide a comprehensive review of the literature of ON and discuss the results.

Chapter 2 - Dietary patterns and food habits during early childhood have long-term impacts in subsequent health outcome in later life. Moreover, parenting style, early feeding practices and child eating behavior are established in early years of life. Growing body of evidence has documented that a healthy dietary pattern is associated with lower risk of the development of non-communicable diseases (NCDs) including diabetes, cardiovascular disease, and some cancers. Early feeding practices of parents and caregivers would determine the type, amount and frequency of foods of their children as well as various eating disorders. These feeding practices are established by five years of age, and would strongly affect eating patterns in childhood and in adulthood. The degree of parental control including restriction, monitoring and pressure over early feeding might have strong impacts on the preferences and intake of healthy or unhealthy foods of children. Unhealthy food intake with nutrient deficiency and poor dietary variety in early life is an important health concern with adverse early- and late consequences for children. Prolonged unhealthy diet can lead to growth failure as well as delays in cognitive and developmental issues. Nutrition is a major modifiable factor related to incidence of chronic diseases. Intake of healthy foods in early life may not only influence current health, but may also be a determinant of the development and progress of NCDs much later in life. This chapter aims to summarize the current literature on the effect of healthy foods and early feeding practices in childhood on primordial prevention of NCDs and their risk factors.

Chapter 3 - Carrot bagasse flour and banana peel flour have the potential to be employed as functional ingredients to improve texture, color, and flavor of raw meat products, as chorizo, or cooked meat products, as sausages. Total dietary fiber content was 40.87% and 31.67% for carrot bagasse flour and banana peel flour, respectively. The natural antioxidants content as total polyphenols were high in carrot bagasse (206 vs. 42 mg catechin/100 g flour), but the antioxidant capacity was high for banana peel flour (283 vs. 191 TEAC). In the same manner, prebiotic activity was higher for carrot bagasse flour (0.63 vs. 0.10). The composition of the flour made them a good candidate for functional meat products extenders, since incorporation in 2% and 4% in a raw meat product as chorizo, and a cooked meat product as sausage, improved yield, where the fiber hydration during process retained more water, improving expressible moisture as well. In sensory acceptation, the general appearance of banana peel flour chorizo was rejected by the consumers, whereas carrot bagasse flour chorizo was more accepted. Sausage texture was more accepted by the consumers for both flours. For cooked sausages' general appearance, carrot bagasse tendency was in the "nor like or dislike", with a better acceptance for banana peel flour samples. Nonetheless, texture has a good acceptation for both flours. However, in the neophobia test, 56% of consumers tend to reject new foods. Still, carrot bagasse flour and banana peel flour can be considered as a good option to improve the nutritional profile of raw or cooked meat products with fiber and antioxidants, with no major effect of texture.

Chapter 4 - Fat in animal food products represents a source of undesirable compounds like saturated fats, associated to worsen cardiovascular disease and obesity. Fat is an important component in emulsified cooked meat products, like sausages. Restructuring oils have become a techno-functional alternative to incorporate polyunsaturated oils as oleogel, improving texture and oxidative stability. The effect of candelilla wax oleogel to replace pork backfat lard was determined on cooked sausages textural profile analysis, moisture, color, and sensory acceptance. Sausages formulated with oleogel resulted harder and more resilient but less cohesive than samples with lard. Oleogel increased total moisture and enhance water retention. The incorporation of oleogel to replace fat resulted in a lighter coloration, with higher tone and saturation index values, and acceptable color difference (<6). As expected, oleogel samples obtained lower oxidative rancidity than lard containing samples. In sensory acceptance, no difference between oleogel sample and control with lard was detected for color, taste, fat sensation, texture or overall acceptance. These results indicate that candelilla wax oleogel can be employed as a fat replacer in emulsified cooked sausage to reduce saturated fats and increase polyunsaturated oils, improving the health profile of this kind of meat products.

In: Healthy Food
Editor: Anthony E. Walton
ISBN: 978-1-53617-599-8
© 2020 Nova Science Publishers, Inc.

Chapter 1

WHAT WE KNOW SO FAR ABOUT ORTHOREXIA NERVOSA: A REVIEW

Márton Kiss-Leizer[1],* and Adrien Rigó[2], PhD
[1]Department of Clinical Psychology,
Semmelweis University, Budapest, Hungary
[2]Department of Personality and Health Psychology,
Eötvös Loránd University, Budapest, Hungary

1. INTRODUCTION

There is an emerging focus on healthy eating in Western culture. Although healthy eating is undoubtedly a core element of healthy life, this behavior can also become excessive and harmful. Being obsessed with healthy eating may lead to eating disorders and, paradoxically, the aspiration to be healthy becomes unhealthy. This phenomenon well represents how eating disorders are influenced by the actual social-cultural environment; recently the ideal of being healthy has gained strength besides the ideal of being thin, which may have contributed to the

---

* Corresponding Author's E-mail: leizermarton@gmail.com.

emergence of the obsessive type of healthy eating (Varga, Dukay-Szabo and Túry, 2013). The pathological way of healthy eating – orthorexia nervosa (ON) - has been described by Bratman (1997) and due to the relevance and actuality of the topic, the number of scientific publications are increasing in this area. Although ON cannot be found in the Diagnostic and Statistical Manual of the American Psychiatric Association (DSM-5) and it is still not fully understood, important findings have suggested that ON should get a wider scientific and public attention. The aim of this paper is to provide a comprehensive review of the literature of ON and discuss the results.

2. METHOD

The keywords "orthorexia nervosa," "healthy eating disorder," "healthy eating pathology" and "healthy food dependence" were searched in several online databases (Google Scholar, Pubmed and PsychINFO). The results (articles, books and book chapters) have been reviewed, but the commentary papers and the unavailable articles via academic library were excluded.

3. RESULTS

3.1. Definition and Criteria

ON is characterized by an excessive preoccupation with healthy food (Bratman, 1997). Orthorexic patients are obsessed with their self-defined dietary restrictions, they are concerned with the quality of food and they consume only what they find to be "pure" (controlled origin, no artificial ingredients, preservatives or additives) (Chaki, Pal and Bandyopadhyay, 2013). In healthy eating addiction the enjoyment and the taste of the food are losing their significance, because only the intake of valuable nutrients is considered when selecting food (Bratman, 1997). Thus, although ON is closely related to the well-known eating disorders (anorexia nervosa, bulimia nervosa), it can obviously be distinguished from them since the focus in anorexia and bulimia is primarily on the amount of food consumed (Larsen, 2013). The phenomenon often begins innocently: the original aim is to overcome a chronic disease or to improve general health. Later the eating habits originating from childhood and the surrounding cultural environment slowly disappear, and are gradually replaced with the tendency to have the quality of food as the main organizing factor of eating, and the issue of what to consume starts to have an increasing impact on everyday life (Bratman & Knight, 2000). The strict dietary habits, in the second stage, are combined with an emerging obsessive compulsive thinking, compulsive behavior, and self-punishment (Bratman, 2017). The act of eating purely might gain a sense of pseudospiritual significance, according to breaking the rules can result in fasting and starving in order to reach "balance" again. It can be also observed that orthorexic people are trying to cope and overcome disappointments in life through pure nutrition (Bratman, 1997).

The term "orthorexia" is a compound of two Greek words: "ortho," meaning straight or correct, and "rexis," which means appetite. The excellent quality of food counts above everything and thus, the severe dietary restrictions can lead to adverse health effects (Dunn & Bratman, 2016). In fact, the medical complications may be quite similar to those observed in chronic anorexic patients: decreased bone density (osteopenia), anemia, salt deficiency, increased bicarbonate loss (metabolic acidosis), decreased red blood cell, white blood cell and platelets counts (pancytopenia), testosterone deficiency, and slow heart rate (bradycardia) (Bratman & Knight, 2000; Moroze, Dunn, Holland, Yager and Weintraub, 2015; Park et al., 2011). Orthorexic patients spend considerable time investigating the source of food, as well as monitoring the preparation procedure, and they eat only in scheduled times, performing eating rituals. Individuals with ON feel extreme guilt and anxiety when they do not eat healthy; intense frustration appears if something disrupts their habits and they are disgusted when the quality of food is inappropriate according to their standards (Dunn & Bratman, 2016). It can be observed that ON patients plan their diet on a compulsive basis, and although obsessive-compulsive disorder (OCD) is a distinct diagnostic category, the overlap between the two disorders is undoubted (Koven & Wabry, 2015). The strict self-discipline is interestingly accompanied with feeling superiority over other eating habits, which can result in separation from others and social isolation (Dunn & Bratman, 2016). According to the diagnostic criteria of orthorexia nervosa (Dunn & Bratman, 2016), health-dependent addictions are characterized by the following:

Criterion A: The obsessive focus of healthy eating can be observed by the following:

- Compulsive behavioral attitudes and/or mental preoccupation in connection with strict eating habits, which are believed by the individual to promote health.
- Increased fear of illness, feeling personal impurity, shame and guilt due to violations of the self-defined eating habits.
- The dietary restriction escalates over time, which may result in rejecting whole groups of foods. The strict diet usually leads to weight loss, but the desire to be thin is not the main target, rather the idealization of being healthy.

Criterion B: The obsessive behavior and the mental disease are accompanied by any of the following:

- Malnutrition, severe weight loss or other health consequences due to the diet.
- Intrapersonal distress or social problems in different areas (everyday life, workplace, academic life).
- Positive body image, high self-esteem or/and feeling satisfied with the self- defined "healthy" eating habits.

3.2. Differential Diagnosis

No research has investigated the differential diagnosis of ON in the literature so far, but there are numerous case studies in this area that help to distinguish between ON and other mental diseases (Varga, Dukay-Szabó, Túry and Van Furth Eric, 2013). One of them (Zamora, Bonaechea, Garcia Sánchez and Rial, 2005) reported a 28-year-old orthorexic patient with 10.7 BMI and with a weight of 27 kg, who could easily have been diagnosed with anorexia nervosa - based on objective data -, but there was no distortion in body image or strong desire for slimness. The obsessive thoughts were rather built around nutritional supplements that are considered to be healthy.

It is also essential to distinguish between health conscious people and those with ON. The main difference between the two groups is that people with ON feel extreme anxiety in connection with food, they consume only healthy food, they judge those who do not eat according to the orthorexic principles and they feel guilty when they cannot keep their prescribed diet (Varga, Dukay-Szabó, et al., 2013).

If the symptomatology of ON can be justified by bizarre, paranoid delusions, ON must be separated from psychotic diseases and schizophrenia. However, in these cases, ON can be described as a symptom of the prodromal stage of schizophrenia (Dunn & Bratman, 2016).

3.3. Epidemiological Data

Epidemiological data of ON are largely influenced by the sensitivity and specificity of the used questionnaires. To date there is no adequately controlled psychometric measuring instrument of ON. In the first article about ON Bratman (1997) presented a dichotomous evaluation scale (Bratman Orthorexia Test - BOT). The BOT is based on the criteria of the author; two or three "yes" answers indicate tendency to ON, whereas four or more "yes" answers mean a danger of ON. The majority of ON studies are based on the ORTO-15 questionnaire, which also measures the tendency to ON; nevertheless, the reliability and the validity of the test are uncertain (Dunn & Bratman, 2016).

One of the first studies in this field was an Italian research project (Donini, Marsili, Graziani, Imbriale and Cannella, 2004), in which the diagnosis of ON was based on OCD symptoms and exaggerated healthy eating behavioral patterns. According to the results, 28 of 404 participants met the criteria of ON, and the prevalence rate was 6.9%. Similarly, relatively low prevalence values were found in a recent American study (Bundros, Clifford, Silliman and Neyman Morris, 2016) with BOT. Interestingly, only 5 of the 448 subjects were affected by ON but according to the authors, the tendency of ON was high in the sample.

The ORTO-15 is a 4-degree Likert scale, based on BOT, supplemented with items about OCD tendencies (Donini, Marsili, Graziani, Imbriale and Cannella, 2005). According to an Italian study (Ramacciotti et al., 2011) 57,6% of the adult population can be featured with tendency to ON. In addition, the research suggested a difference between men and women with the ratio of 2:1. A considerably high rate was also confirmed by US research (Dunn, Gibbs, Whitney and Starosta, 2017) with 72%. These are somewhat contradicted by an Italian study (Cinosi et al., 2015), which targeted mainly young people and found a prevalence rate of "only" 10,9%. However, higher values on ORTO-15 were confirmed in women. A Hungarian study (Varga, Thege, Dukay-Szabó, Túry and van Furth, 2014) found a similarly high prevalence rate of tendency to ON among a sample of undergraduate students (about 74.2%) with the Hungarian version of ORTO-15 (Orto-11-Hu). The results were independent of gender and BMI. The authors suggested that the cut-off point of the test should be revised; and based on their results a lower cut-off point has been defined. The absence of the effect of BMI was confirmed by another study with university students (Sanlier, Yassibas, Bilici, Sahin and Celik, 2016), but it showed a possible influence of gender, with a higher risk to ON in women.

In the light of the relatively high prevalence results with ORTO-15 a new, different measurement method of ON has been published: the Eating Habits Questionnaire (EHQ) (Oberle, Samaghabadi and Hughes, 2017).

The EHQ was developed independently from ORTO-15, and the final 21-item version showed promising reliability and validity values. According to the authors' results - in line with most of the previous findings - there were no significant gender differences in ON symptomatology, although men reported greater healthy eating behavior, whereas women reported greater positive feeling in connection with eating healthy. In this study high BMI predicted more severe ON symptoms, but only in men.

The limited prevalence data regarding ON are dominated by studies using ORTO-15 or its adaptations, but the results are quite diverse and inconsistent; consequently it is hard to make clear conclusions about the frequency of ON in the population and the possible effect of different variables (BMI, gender).

3.4. Risk Groups in Connection With ON

Certain groups have been identified to have an increased risk of ON, which is particularly important in planning prevention and intervention programs. In order to increase sports performance, many athletes cultivate strictly regulated eating habits that can lead to different eating disorders. In line with that, the athletes' scores on ORTO-15 were significantly higher than in the control group (Segura-García et al., 2012), which drew attention to the dysfunctional eating habits. The higher risk for ON in people characterized by more frequent physical activity was confirmed by another study (Stochel, Hyrnik, Jelonek, Zejda and Janas-Kozik, 2013), which also shed light on the correlation between higher BMI and higher ON tendency.

Performers are often in a model role for the general population and thus they can be particularly sensitive towards their appearance. Examining various types performance artists (Emine Aksoydan & Camci, 2009), the highest prevalence ON rate was identified among opera singers but a high rate among symphonic musicians and ballet dancers also emerged. The result was not correlated with gender, age, qualification, BMI, work experience or typical health behaviors (smoking or drinking).

Since doctors and medical students are possibly in an exemplary role for the general population in terms of maintaining health, they are more likely to be affected by ON (Bağci Bosi, Çamur and Güler, 2007; Fidan, Ertekin, Işikay and Kirpinar, 2010). The high scores on ORTO-15 in the groups of physician and medical students went hand in hand with purchasing higher quality foods, weight control behaviors and with shopping for food alone. Similarly, high prevalence rates have been observed among nutrition students and dieticians (Alvarenga et al., 2012). Additionally, there was a positive correlation between healthy eating habits and nutritional knowledge, which could explain dietitians' high risk of pathological ways of healthy eating (Asil & Sürücüoğlu, 2015). Compared with a control group, dietitian students were characterized by stricter dietary rules and a more rigid control of eating (Varga, Dukay-Szabo, et al., 2013). Further investigation found that 12.8% of dietetic students (Kinzl, Hauer, Traweger and Kiefer, 2006) were particularly vulnerable to ON, based on Bratman's criteria, and nearly 5% reported eating disorders (anorexia, bulimia or binge eating disorder). As these are cross-sectional studies, we can't see the underlying mechanisms; it is possible that those people are choosing to study dietetics who are a priori vulnerable to eating disorders; or studying about diet makes people more prone to developing these disorders.

In many cases, ON masks the "classical" eating disorders; it may be a preliminary phase of later anorexia, but can also act as a solution mechanism for overcoming other eating disorders (Varga, Dukay-Szabo, et al., 2013). Assessing anorexic and bulimic patients for ON and for various eating attitudes before and after a three-year treatment, helped to shed light on the connection between AN, BN and ON (Segura-Garcia et al., 2015). In the analysis the eating disorder groups were matched with a healthy control group in terms of age, BMI and gender. The results showed that the prevalence of ON began to increase after the treatment. Moreover, ON was not only associated with clinical improvement, but also with the decrease of the severity of the other eating disorders. Nevertheless, it is important to clarify that the residual ON symptoms can be in connection with a higher risk of relapse and reoccurrence of eating disorders.

There are people who start to follow a specific diet for medical reasons, in which the high quality and pure food appear to be the only right choice. In this regard, it is worth mentioning that patients are increasingly receiving dietary recommendations from doctors for different problems (asthma, IBS, allergies) to eliminate the irritable components from their diet, but, in some cases, these are not well justified and can lead to fixations in nutrition (Arhire, 2015). Investigating the connections to alternative foods in people with ON and in people who follow special diets (vegetarian, vegan, paleo, gluten-free, etc.), it was found that people with special diets are more likely to be affected by ON or even by other eating disorders (Barnett, Dripps and Blomquist, 2016).

Yoga, along with healthy lifestyle is gaining ground in Western societies, and has beneficial effects not only on physical, but also on mental and spiritual well-being. Despite the positive influence, Asthanga yoga communities were found to have a higher risk of developing ON. According to ORTO-15 scores, 86% of the 136 participants have high tendency to ON regardless of BMI or age (Herranz Valera, Acuña Ruiz, Romero Valdespino and Visioli, 2014). The research has also confirmed that vegetarian population is indeed a risk group.

In summary, research has successfully identified a number of risk groups of ON: doctors (Bağcı Bosi, Çamur and Güler, 2007) or medical students (Fidan et al., 2010), dietitians (Asil & Sürücüoğlu, 2015), dietitian students (Alvarenga et al., 2012; Kinzl et al., 2006), athletes (Segura-García et al., 2012), performance artists (E. Aksoydan & Camci, 2009), or regular practitioners of asthanga yoga (Herranz Valera et al., 2014). The results are controversial in terms of gender and BMI. Some studies found no association between BMI and ON, while others stated that high BMI - which indicates obesity - is correlated with high ON tendency. The majority of the studies found no significant differences between males and females in terms of ON tendencies, while a few found a higher risk for ON in women. In contrast, in a study (Donini et al., 2004) ON was higher in male participants.

3.5. Personality Features of ON

In light of the spectrum disorder approach, it is worth considering the relevant overlap between the characteristics of the "classical" eating disorders (anorexia, bulimia) and ON. It is possible, that they are in the same psychopathological spectrum, but in ON, the focus is on the quality of the food, not on the quantity. In another point of view, anorexic and bulimic patients are seeking a socially accepted way of weight control, while people with ON mostly focus on the healthiness/pureness of food, not on their body shapes. In order to explore the characteristics of ON, participants were investigated in terms of their ON tendencies, perfectionism, body image and attachment styles (Barnes & Caltabiano, 2017). Analysis revealed that high ON tendencies correlated with every scale of multidimensional perfectionism (against themselves, against others and against society), with orientation towards appearance, with overweight preoccupation and fearful/avoidant attachment styles. It is important to note, that these features have also been observed in anorexia and bulimia. In addition, earlier eating disorders were strong explanatory factors for later ON. The possible role of perfectionism in the etiology of ON has been confirmed by other research, which describes ON patients as cautious, meticulous, neat people with an increased demand of self-care and protection, along with strong addictive impulses (Fidan et al., 2010). Moreover, recent research found that ON is not only associated with perfectionism, but also with narcissism (Oberle et al., 2017).

The overlap between ON and OCD tendencies has been described in the first article about ON (Bratman, 1997), and has been confirmed later by others. Examining the relationship between various anxiety disorders (OCD, panic disorder, generalized anxiety disorder) and ON (Poyraz et al., 2016), there were no significant differences detected between patient groups in terms of eating habits and ON severity. However, moderate correlation has been observed between ON symptoms severity and OCD symptoms severity, eating attitudes and controlling/dressing compulsive mechanisms. The results showed that although pathological healthy eating is not specifically related to these mental illnesses, it can have a moderator effect on the ritual behavior of OCD.

Further similarities have been documented between the two disorders. As in OCD, ON patients also have recurring, intrusive thoughts about eating and health, as well as increased concern about infections, contaminations and impurity. ON people – like OCD people – have reduced free time due to their activities as a result of their disease, and their strict eating style makes them difficult to practice every-day routines. However, an important difference is that while the symptoms in OCD are manifested in an egodiston (behaviors, values, feelings that are inconsistent, dissonant with the person's self-image) form, they are egosynton (behaviors, values, feeling that are consistent with the person's self-image) in the case of ON (Koven & Wabry, 2015).

ON and the restrictive subtype of anorexia share many characteristics that are primarily classified as OCD features. The most specific traits are rigidity, a strong need for control and perfectionism, which are manifested in healthy eating (Brytek-Matera, 2012). ON patients are characterized by emotional eating, which means that their happy, sad, guilty or excited mood affects their eating habits, for instance they feel uncontrollable desire to eat either when they are in a positive mood or nervous. In the case of ON, immense anxiety appears around eating only pure food, which includes specified preparation procedures, avoiding the consumption of certain foods and a high level of hypochondriac symptoms (Brytek-Matera, 2012). The strong need for control and the pathological desire for perfection is confirmed by another study (Mathieu, 2005). These features are associated with certain cognitive difficulties, which indicate that people with ON have weaknesses in certain executive functions, such as switching attention, emotional control, capacity of working memory and self-monitoring functions (Koven & Senbonmatsu, 2013).

It seems obvious that healthy food addiction is associated with anxiety disorders, so it is not surprising that the tendency to ON is related to high social anxiety (Eriksson, Baigi, Marklund and Lindgren, 2008). Apparently, it is connected with the internalization of socio-cultural attitudes: it has been found that openness, awareness and internalization of

Western cultural values about physical fitness and slimness increase the vulnerability to ON. Moreover, internalization itself could explain the variability of BOT results, independently from gender. In women more frequent exercise was associated with higher tendency to ON, while the internalization of ideal muscular body showed similar results in men ((Eriksson et al., 2008). In addition, ON patients, such as anorexic patients, are characterized by a high level of trait anxiety and need to exert control (Koven & Wabry, 2015).

A recent study (Kiss-Leizer & Rigó, 2018) investigated the personality profile of people with high tendency to ON with a measurement method which is widely used in the research of classical eating disorders (anorexia, bulimia) and OCD. The results revealed that high ON tendencies are associated with higher harm avoidance, lower self-directness and higher transcendence in terms of Temperament and Character Inventory (TCI). The authors suggested that high harm avoidance and low self-directness are important factors of anorexia, bulimia and OCD, but the higher transcendence scores are unique to ON. Therefore the excessive preoccupation of healthy food is not just a lifestyle, but rather a personal philosophy, which creates an illusion of safety for the orthorexic individual. This can be paralleled with Bratman's original statements of ON, in which he described orthorexic patients, as people who often try to solve their problems and disappointment through healthy eating and they form specific spiritual rituals around the preparation procedure of food.

According to the cited studies, the concept of ON can be described as a complex, multilevel disorder, which contains the personality characteristics of eating, behavior and OCD disorders (Donini et al., 2004). However, it is important to highlight that ON retains its autonomy and is clearly distinguishable from other mental illnesses. That explains why some authors emphasize that the symptoms of healthy food addiction cannot be described as episodes of other psychiatric diseases (such as OCD, anorexia, psychosis or hypochondria) (Varga, Dukay-Szabo, et al., 2013).

4. Discussion

Summarizing the literature, it can be stated that dynamically improving and increasing interest can be observed in the non-professional and professional communities towards the topic of orthorexia nervosa. Besides the obesogenic environment and the ideal of slimness, the ideal of healthiness emerged, which resulted in the dietary habits of people becoming more and more complex. The rapid expansion of the concept of ON is explicable by the fact that in the society of the 21st century numerous people are affected by the symptoms. Orthorexia nervosa describes the phenomenon well, although, the literature still raises many questions. Research successfully identified possible risk groups, background factors, personality features and relationships with other psychiatric disorders, but the prevalence data are contradictory, and the place of the disease in the diagnostic and nosological systems is still questionable. The development and the course of ON are influenced by a number of factors, thus a multidimensional, holistic perspective is essential to comprehend the phenomenon. However, more accurate knowledge is required about the nature of the disorder and about the characteristics of the individuals in order to scientifically approve ON as a mental disorder.

It can be seen that the factors and mechanisms that can lead to ON are very diverse, because healthy eating can be motivated by social benefits, achieving the ideal body, fear of diseases, overcoming of a chronic illness or solving everyday problems through pure dietary habits. Future studies should investigate the possible ON subtypes, based on the individuals' background factors, eating patterns and personality parameters. In addition, it would be beneficial to describe the need for pure nutrition as a dimensional construct. Since the phenomenon itself relates to a healthy way of life, in certain mild cases it may have positive or neutral impacts on quality of life, while crossing a line can lead to mental and behavioral problems.

Avoidance of emotions and affect dysregulation play a role in pathological eating habits and are common features of anorexia nervosa and bulimia nervosa (Lavender et al., 2015). It has been widely documented that emotional regulation plays a mediator role in the relationship between attachment and different forms of eating pathology (Durme, Braet and Goossens, 2015) and in depression (Malik, Wells and Wittkowski, 2015). Alexithymia is also strongly related to eating disorders, since the difficulties of understanding and expressing emotions can manifest in eating pathology (Brewer, Cook, Cardi, Treasure and Bird, 2015). It would be beneficial to investigate the possible role of emotional regulation, alexithymia and coping mechanisms in the etiology of ON in the future as well, since ON and the "classical" eating disorders share many characteristics (Brytek-Matera, 2012; Kiss-Leizer & Rigó, 2018), so special emotional functioning might be a core feature of ON too. In parallel, it would be useful to assess whether the construct of high emotional intelligence can have a possible protective factor against the development of such behavior.

It would also be beneficial to examine the possible role of health anxiety in the etiology of ON, because it has been documented that hypochondriac symptoms (Brytek-Matera, 2012) and high levels of trait anxiety (Koven & Wabry, 2015) go together with ON symptoms. In the light of the possibly high level of health anxiety in ON, future studies should also investigate the association between ON and interoceptive awareness, because health anxious people often misinterpret and exaggerate their body signals (Alberts, Hadjistavropoulos, Jones and Sharpe, 2013).

REFERENCES

Aksoydan, E. and Camci, N. (2009). Prevalence of orthorexia nervosa among Turkish performance artists. *Eating and Weight Disorders*, *14*(1), 33–37. https://doi.org/6158 [pii].

Aksoydan, E. and Camci, N. (2009). Prevalence of orthorexia nervosa among Turkish performance artists. *Eating and Weight Disorders -*

Studies on Anorexia, Bulimia and Obesity, *14*(1), 33–37. https://doi.org/10.1007/BF03327792.

Alberts, N. M., Hadjistavropoulos, H. D., Jones, S. L. and Sharpe, D. (2013). The Short Health Anxiety Inventory: A systematic review and meta-analysis. *Journal of Anxiety Disorders*, *27*(1), 68–78. https://doi.org/10.1016/j.janxdis.2012.10.009.

Alvarenga, M. D. S., Martins, M. C. T., Sato, K. S. C. J., Vargas, S. V. A, Philippi, S. T. and Scagliusi, F. B. (2012). Orthorexia nervosa behavior in a sample of Brazilian dietitians assessed by the Portuguese version of ORTO-15. *Eating and Weight Disorders*, *17*(1), 29–35. https://doi.org/10.1007/BF03325325.

Arhire, L. I. (2015). Orthorexia Nervosa : The unhealthy obssesion for healthy food. *Revista Medico-Chiruricala A Societatii de Medici Si Naturalisti Din Iasi*, *119*(3), 632–638.

Asil, E. and Sürücüoğlu, M. S. (2015). Orthorexia Nervosa in Turkish Dietitians. *Ecology of Food and Nutrition*, *54*(4), 1–11. https://doi.org/10.1080/03670244.2014.987920.

Bağcı Bosi, A. T., Çamur, D. and Güler, Ç. (2007). Prevalence of orthorexia nervosa in resident medical doctors in the faculty of medicine (Ankara, Turkey). *Appetite*, *49*(3), 661–666. https://doi.org/10.1016/j.appet.2007.04.007.

Barnes, M. A. and Caltabiano, M. L. (2017). The interrelationship between orthorexia nervosa, perfectionism, body image and attachment style. *Eating and Weight Disorders*, *22*(1), 177–184. https://doi.org/10.1007/s40519-016-0280-x.

Barnett, M. J., Dripps, W. R. and Blomquist, K. K. (2016). Organivore or organorexic? Examining the relationship between alternative food network engagement, disordered eating, and special diets. *Appetite*, *105*, 713–720. https://doi.org/10.1016/j.appet.2016.07.008.

Bratman, S. (1997). Healthy Food Junkie: Obsession with dietary perfection can sometimes do more harm than good, says one who has been there. *Yoga Journal*, *136*, 42–46.

Bratman, S. (2017). Orthorexia vs. theories of healthy eating. *Eating and Weight Disorders - Studies on Anorexia, Bulimia and Obesity*, 22(3), 381–385. https://doi.org/10.1007/s40519-017-0417-6.

Bratman, S. and Knight, D. (2000). *Health Food Junkies: Orthorexia Nervosa - Overcoming the Obsession with Healthful Eating*. New York: Broadway Books.

Brewer, R., Cook, R., Cardi, V., Treasure, J. and Bird, G. (2015). Emotion recognition deficits in eating disorders are explained by co-occurring alexithymia. *Royal Society Open Science*, 2(140382), 1–12. https://doi.org/doi:10.1098/rsos.140382.

Brytek-Matera, A. (2012). Orthorexia nervosa-an eating disorder, obsessive-compulsive disorder or disturbed eating habit? *Archives of Psychiatry and Psychotherapy*, 1, 55–60. Retrieved from http://www.ncbi.nlm.nih.gov/pubmed/22361450.

Bundros, J., Clifford, D., Silliman, K. and Neyman Morris, M. (2016). Prevalence of Orthorexia nervosa among college students based on Bratman's test and associated tendencies. *Appetite*, 101, 86–94. https://doi.org/10.1016/j.appet.2016.02.144.

Chaki, B., Pal, S. and Bandyopadhyay, A. (2013). Exploring scientific legitimacy of orthorexia nervosa: A newly emerging eating disorder. *Journal of Human Sport and Exercise*, 8(4), 1045–1053. https://doi.org/10.4100/jhse.2013.84.14.

Cinosi, E., Matarazzo, I., Marini, S., Acciavatti, T., Lupi, M., Corbo, M., … Di Giannantonio, M. (2015). Prevalence of orthorexia nervosa in a population of young Italian adults. *European Psychiatry*, 30, 13–30. https://doi.org/10.1016/S0924-9338(15)31038-5.

Donini, L. M., Marsili, D., Graziani, M. P., Imbriale, M. and Cannella, C. (2004). Orthorexia nervosa: A preliminary study with a proposal for diagnosis and an attempt to measure the dimension of the phenomenon. *Eating and Weight Disorders*, 9(2), 151–157. https://doi.org/10.1007/BF03325060.

Donini, L. M., Marsili, D., Graziani, M. P., Imbriale, M. and Cannella, C. (2005). Orthorexia nervosa: Validation of a diagnosis questionnaire. *Eating and Weight Disorders*, 10, 28–32.

Dunn, T. M. and Bratman, S. (2016). On orthorexia nervosa: A review of the literature and proposed diagnostic criteria. *Eating Behaviors*, *21*, 11–17. https://doi.org/10.1016/j.eatbeh.2015.12.006.

Dunn, T. M., Gibbs, J., Whitney, N. and Starosta, A. (2017). Prevalence of orthorexia nervosa is less than 1%: data from a US sample. *Eating and Weight Disorders*, *22*(1), 1–8. https://doi.org/10.1007/s40519-016-0258-8.

Durme, K. van, Braet, C. and Goossens, L. (2015). Insecure Attachment and Eating Pathology in Early Adolescence: Role of Emotion Regulation. *Journal of Early Adolesence*, *35*(1), 54–78. https://doi.org/10.1177/0272431614523130.

Eriksson, L., Baigi, A., Marklund, B., & Lindgren, E. C. (2008). Social physique anxiety and sociocultural attitudes toward appearance impact on orthorexia test in fitness participants. *Scandinavian Journal of Medicine and Science in Sports*, *18*(3), 389–394. https://doi.org/10.1111/j.1600-0838.2007.00723.x.

Fidan, T., Ertekin, V., Işikay, S. and Kirpinar, I. (2010). Prevalence of orthorexia among medical students in Erzurum, Turkey. *Comprehensive Psychiatry*, *51*(1), 49–54. https://doi.org/10.1016/j.comppsych.2009.03.001.

Herranz Valera, J., Acuña Ruiz, P., Romero Valdespino, B. and Visioli, F. (2014). Prevalence of orthorexia nervosa among ashtanga yoga practitioners: a pilot study. *Eating and Weight Disorders*, *19*(4), 469–472. https://doi.org/10.1007/s40519-014-0131-6.

Kinzl, J. F., Hauer, K., Traweger, C. and Kiefer, I. (2006). Orthorexia nervosa in dieticians. *Psychotherapy and Psychosomatics*, *75*(6), 395–396. https://doi.org/10.1159/000095447.

Kiss-Leizer, M. and Rigó, A. (2018). People behind unhealthy obsession to healthy food: the personality profile of tendency to orthorexia nervosa. *Eating and Weight Disorders - Studies on Anorexia, Bulimia and Obesity*, 1–7. https://doi.org/10.1007/s40519-018-0527-9.

Koven, N. S. and Senbonmatsu, R. (2013). A neuropsychological evaluation of orthorexia nervosa. *Open Journal of Psychiatry*, *3*, 214–222. https://doi.org/10.4236/ojpsych.2013.32019.

Koven, N. S. and Wabry, A. (2015). The clinical basis of orthorexia nervosa: Emerging perspectives. *Neuropsychiatric Disease and Treatment*, *11*, 385–394. https://doi.org/10.2147/NDT.S61665.

Larsen, K. I. (2013). *Similarities and differences between eating disorders and orthorexia nervosa*. Universitá di Roma.

Lavender, J. M., Wonderlich, S. A., Engel, S. G., Gordon, K. H., Kaye, W. H. and Mitchell, J. E. (2015). Dimensions of emotion dysregulation in anorexia nervosa and bulimia nervosa: A conceptual review of the empirical literature. *Clinical Psychology Review*, *40*, 111–122. https://doi.org/10.1016/j.cpr.2015.05.010.

Malik, S., Wells, A. and Wittkowski, A. (2015). Emotion regulation as a mediator in the relationship between attachment and depressive symptomatology : A systematic review. *Journal of Affective Disorders*, *172*, 428–444. https://doi.org/10.1016/j.jad.2014.10.007.

Mathieu, J. (2005). What is orthorexia? *Journal of the American Dietetic Association*, *105*(10), 1510–1512. https://doi.org/10.1016/j.jada.2005.08.021.

Moroze, R. M., Dunn, T., Holland, J. C., Yager, J. and Weintraub, P. (2015). Microthinking About Micronutrients : A Case of Transition From Obsessions About Healthy Eating to Near-Fatal " Orthorexia Nervosa " and Proposed Diagnostic Criteria. *Psychosomatics*, *56*, 397–403. https://doi.org/10.1016/j.psym.2014.03.003.

Oberle, C. D., Samaghabadi, R. O. and Hughes, E. M. (2017). Orthorexia nervosa: Assessment and correlates with gender, BMI, and personality. *Appetite*, *108*(1), 303–310.

Park, S. W., Kim, J. Y., Go, G. J., Jeon, E. S., Pyo, H. J. and Kwon, Y. J. (2011). Orthorexia Nervosa with Hyponatremia, Subcutaneous Emphysema, Pneumomediastimum, Pneumothorax, and Pancytopenia. *Electrolyte Blood Press*, *9*, 32–37. https://doi.org/10.5049/EBP.2011.9.1.32.

Poyraz, C. A., Tüfekçioğlu, E. Y., Özdemir, A., Bas, A., Kani, A. S., Erginöz, E. and Duran, A. (2016). Relationship between Orthorexia and Obsessive-Compulsive Symptoms in patients with Generalised Anxiety Disorder, Panic Disorder and Obsessive Compulsive Disorder. *Yeni Symposium*, *53*(4), 1. https://doi.org/10.5455/NYS.20151 221025259.

Ramacciotti, C. ., Perrone, P., Coli, E., Conversano, C., Massimetti, G. and Dell'Osso, L. (2011). Orthorexia nervosa in the general population : A preliminary screening using a self-administered questionnaire (ORTO-15). *Eating and Weight Disorders*, *16*, 127–130. https://doi.org/10.1007/BF03325318.

Sanlier, N., Yassibas, E., Bilici, S., Sahin, G. and Celik, B. (2016). Does the rise in eating disorders lead to increasing risk of orthorexia nervosa? Correlations with gender, education, and body mass index. *Ecology of Food and Nutrition*, *55*(3), 266–278. https://doi.org/10.1080/03670244.2016.1150276.

Segura-García, C., Papaianni, M. C., Caglioti, F., Procopio, L., Nisticò, C. G., Bombardiere, L., … Capranica, L. (2012). Orthorexia nervosa: A frequent eating disordered behavior in athletes. *Eating and Weight Disorders*, *17*(4), 1–17. https://doi.org/10.3275/8272.

Segura-Garcia, C., Ramacciotti, C., Rania, M., Aloi, M., Caroleo, M., Bruni, A., … De Fazio, P. (2015). The prevalence of orthorexia nervosa among eating disorder patients after treatment. *Eating and Weight Disorders*, *20*(2), 161–166. https://doi.org/10.1007/s40519-014-0171-y.

Stochel, M., Hyrnik, J., Jelonek, I., Zejda, J. and Janas-Kozik, M. (2013). Orthorexia among Polish urban youth. *European Neuropsychopharmacology*, *23*(2), 527–528. https://doi.org/10.1016/S0924-977X(13)70837-X.

Varga, M., Dukay-Szabo, S. and Túry, F. (2013). Orthorexia nervosa and it's background factors. *Ideggyógyászati Szemle*, *66*(7–8), 220–227. Retrieved from http://www.ncbi.nlm.nih.gov/pubmed/23971352.

Varga, M., Dukay-Szabó, S., Túry, F. and Van Furth Eric, F. (2013). Evidence and gaps in the literature on orthorexia nervosa. *Eating and*

Weight Disorders, *18*(2), 103–111. https://doi.org/10.1007/s40519-013-0026-y.

Varga, M., Thege, B. K., Dukay-Szabó, S., Túry, F. and van Furth, E. F. (2014). When eating healthy is not healthy: orthorexia nervosa and its measurement with the ORTO-15 in Hungary. *BMC Psychiatry*, *14*(59), 1–11. https://doi.org/10.1186/1471-244X-14-59.

Zamora, C. M. L., Bonaechea, B. B., Garcia Sánchez, F. and Rial, R. B. (2005). Orthorexia nervosa: A new eating behavior disorder? *Actas Espanolas de Psiquiatria*, *33*(1), 66–68.

In: Healthy Food
Editor: Anthony E. Walton

ISBN: 978-1-53617-599-8
© 2020 Nova Science Publishers, Inc.

Chapter 2

HOW HEALTHY FOODS AND EARLY FEEDING PRACTICES CAN BE EFFECTIVE IN PRIMORDIAL PREVENTION OF NON-COMMUNICABLE DISEASES

Motahar Heidari-Beni[1] and Roya Kelishadi[2],*

[1]Assistant Professor of Nutrition, Department of Nutrition,
Child Growth and Development Research Center,
Research Institute for Primordial Prevention of Non-Communicable
Disease, Isfahan University of Medical Sciences, Isfahan, Iran
[2]Professor of Pediatrics, Department of Pediatrics
Child Growth and Development Research Center,
Research Institute for Primordial Prevention of Non-Communicable
Disease, Isfahan University of Medical Sciences, Isfahan, Iran

---

* Corresponding Author's E-mail: kelishadi@med.mui.ac.ir and roya.kelishadi@gmail.com.

Abstract

Dietary patterns and food habits during early childhood have long-term impacts in subsequent health outcome in later life. Moreover, parenting style, early feeding practices and child eating behavior are established in early years of life.

Growing body of evidence has documented that a healthy dietary pattern is associated with lower risk of the development of non-communicable diseases (NCDs) including diabetes, cardiovascular disease, and some cancers.

Early feeding practices of parents and caregivers would determine the type, amount and frequency of foods of their children as well as various eating disorders. These feeding practices are established by five years of age, and would strongly affect eating patterns in childhood and in adulthood. The degree of parental control including restriction, monitoring and pressure over early feeding might have strong impacts on the preferences and intake of healthy or unhealthy foods of children.

Unhealthy food intake with nutrient deficiency and poor dietary variety in early life is an important health concern with adverse early- and late consequences for children. Prolonged unhealthy diet can lead to growth failure as well as delays in cognitive and developmental issues.

Nutrition is a major modifiable factor related to incidence of chronic diseases. Intake of healthy foods in early life may not only influence current health, but may also be a determinant of the development and progress of NCDs much later in life.

This chapter aims to summarize the current literature on the effect of healthy foods and early feeding practices in childhood on primordial prevention of NCDs and their risk factors.

Keywords: diet, feeding behavior, non-communicable diseases, primary prevention

Parental or Caregivers Influence on Children's Food Intake or Food Preferences

Experienced positive social support and motivation from parents lead to intake healthy foods, modify child eating behaviors and child weight status [1]. The results of family focused study showed that parental education and increasing in nutrition knowledge led to reduction in saturated fat intake in 4-13 year old children [2]. Some parents encourage their children for eating healthy diet and try to limit them for consuming sweets, chocolates, soft drinks and potato crisps. Parents influence on the amount of fruits and vegetables intake of children and adolescents. If the family members eat fruit and vegetables, children learn how to eat them [3, 4].

Study on 698 first-time mothers with healthy term infants showed mothers that received anticipatory guidance on protective feeding practices or usual care had children with higher preference for fruits, higher satiety responsiveness and lower food responsiveness. Child Dietary Questionnaire score for fruit and vegetables was higher in intervention children [1].

According to recently systematic review, there are fewer studies that assess the effect of food parenting includes parent feeding practices and feeding styles on the snacking behaviors of children. Various parenting styles including uninvolved, authoritarian or pressuring child to eat have been associated with different childhood dietary pattern and weight-related outcomes [5].

Results of cross-sectional study showed that obesogenic behaviors of mothers including television (TV) or video watching, fast food intake and sugar-sweetened beverage consumption were strongly associated with these behaviors in their preschool aged children [6].

One cluster randomized clinical trial study assessed the impact of the six months Healthy School Start parental support program on physical activity, dietary habits and body weight in 6 years old children. This intervention was applied in families with low socio-economic status in the school context. Consumption of unhealthy food and drinks were reduced in intervention group that this beneficial effect was sustained for boys after follow-up. 70% of the children consumed at least two servings of fruit and vegetables per day. So, intake of fruit and vegetable remained without changes after intervention. The amount of unhealthy food intake decreased after intervention. A systematic review confirms that educating of parents and increasing their knowledge lead to improve dietary habit of their children [7].

The efficacy of Food, Fun, and Families (FFF) parenting intervention was assessed for 12 weeks in mothers with low-income levels. The aim of the study was reducing consumption of calories from solid fat and added sugar. The results showed the efficacy of the FFF parenting intervention on dietary behaviors. So, nutrition education interventions on parents can have beneficial effects on improving the nutrition of children [8].

Most of the studies focus on the mothers' behaviors. However, behaviors of fathers associated with the development of dietary behaviors and subsequent weight outcomes of children. Since the perceptions of father on foods and dietary behaviors are important, education is needed for the whole family to modify their dietary patterns and prevent obesogenic behaviors [9, 10].

EARLY HEALTHY FOODS AND THE RISK OF DIABETES

One of the most common autoimmune diseases is type I diabetes that develop during infancy. Destruction of β-cells within the pancreas occur in especially genetically susceptible individuals. Certain nutrients and possible toxic food components develop the autoimmune β-cell destruction [11].

Healthy nutrition is recommended for prevention and treatment of type I diabetes. The intake of simple carbohydrates, saturated fatty acids, trans fatty acids isomers, and table salt should be decreased and the consumption of complex carbohydrates (including dietary fiber) and unsaturated fatty acids should be increased. Suitable dietary habits lead to normoglycemia, normal body weight, normal serum lipid levels and normal blood pressure for prevention of complications of diabetes [12].

The frequency of foods rich in protein, carbohydrates (mono- and disaccharides), and nitrosamines intake associated with the risk of developing childhood diabetes. Study on 7-14 year old children showed that disaccharides and sucrose was correlated with increasing the risk of diabetes [11].

Elevated BMI correlated with insulin resistance and type II diabetes. Childhood BMI and systolic blood pressure predict type II diabetes and insulin resistance in young adult life (13). Cohort study on children with 20 years follow up found that 60% of 9–11 years old children in the highest BMI quartile were obese in young adulthood. Obesity in childhood increases three time risk of diabetes in adulthood [14].

The gut microbiota through short-chain fatty acid (SCFA) impact on the development of type I diabetes. Nutritional intervention with increasing SCFA production may be considered as a novel prevention strategy. SCFA are produced in the colon after bacterial fermentation of non-digestible dietary carbohydrates. Acetate, propionate, and butyrate are the main SCFA. They influence on gut integrity and have anti-inflammatory effects through signaling path-ways such as activating G-protein-coupled receptors, inhibiting histone deacetylase, stimulation of histone acetyl transferase activity, and stabilizing hypoxia inducible factor. Some evidence showed the protective effect of SCFAs against type I diabetes [15, 16].

According to animal studies, feeding the mice with acetate or butyrate releasing diets increased levels of acetate and butyrate in the feces, in hepatic and peripheral blood. They showed strong opposite association between SCFAs levels and progression to diabetes. SCFAs can control the development of autoimmune diseases such as autoimmune diabetes.

Acetate can inhibit the proliferation of autoimmune T effector cells by reducing the number of splenic B-cells. They have a pathogenic role in autoantibody production and in the transition from insulitis to clinical diabetes. In addition, they serve as important antigen-presenting cells for islet antigen-reactive T-cells [17].

EARLY HEALTHY FOODS AND THE LONG-TERM RISK OF CARDIOVASCULAR DISEASES (CVD)

Lifestyle interventions in childhood for the prevention and treatment of CVD risk factors are important because of the early nature of the development of CVD. Nutrition and healthy dietary pattern in childhood and adolescence play an important role in the development of CVD in adulthood [18].

Unhealthy dietary pattern, unhealthy dietary nutrient intake patterns and takeaway meal consumption correlated with higher intake of energy, fat, saturated fat and higher energy density and lower protein and micronutrient intakes. Unhealthy foods associated with higher fat mass index and dyslipidemia including higher serum total cholesterol and low-density lipoprotein (LDL) cholesterol, obesity and coronary heart disease risk. Eating a takeaway meal equal or more than one time per week lead to increase 0.09 mmol/L of total cholesterol and 0.10 mmol/L of LDL cholesterol compare with never/hardly ever eating a takeaway meal [19].

Expert Panel on Integrated Guidelines for Cardiovascular Health and Risk Reduction in Children and Adolescents suggest specific dietary intervention for reduction CVD risk. The total fat intake should be limited to 30% of daily calories, saturated fat intake limited to 7-10%, MUFA 10% and polyunsaturated fatty acids (PUFA) 10% of calories and dietary cholesterol limited to 200–300 mg/day. Further trans fatty acids must be avoided. Protein intakes should account for 15-20% and carbohydrate intakes should account for 50-55% of daily calories. These recommendations can be used for children older than 2 years [20, 21].

Dietary habits develop from early life. Healthy dietary habits need for optimal growth and development. In addition, healthy dietary pattern is essential for prevention of cardiovascular risk factors from childhood to adulthood. Some dietary items including calorie intake, the type of fat intake, salt intake and adequate amount of vegetables and fruits have important impact on the development and the progress of cardiovascular disease [22, 23].

Rapid growth because of higher intake of calories and unhealthy diet particularly during early childhood as critical periods, may associate with obesity development and CVD risk factors later in life. Rapid growth between the ages of 2-11 years is correlated with higher risk of obesity and cardio-metabolic diseases, which have adverse health effects. So, growth monitoring in childhood may be beneficial for decreasing long-term health effects [24].

Study on 1000 infant 7 months aged, showed that low-saturated-fat and low cholesterol diet that was limited to total fat 30-35% of calories, saturated fat/MUFA+PUFA ratio of 1:2 and cholesterol intake < 200 mg/day led to better lipid profile in childhood and adolescence [25]. Another study reported that restriction of saturated fat from infancy correlated with lower systolic and diastolic blood pressure in childhood and adulthood [26].

Low sugar-sweetened beverage intake, limited sodium intake, high dietary fiber intake and intake of whole grain, nuts, legumes, fish and virgin olive oil should be promoted since childhood and adolescence for prevention of CVD risk later in life [27].

Hypertension and dyslipidemia are the major risk factors for CVD. Hypertension and dyslipidemia in childhood are correlated with these disorders in adulthood. Decreased intake of sodium and reduced consumption of dietary fats, cholesterol, saturated fat, and trans fat are dietary interventions for hypertension and dyslipidemia respectively. Recently, findings showed that added sugars are associated with hypertension and dyslipidemia. Sugar-sweetened beverages (SSBs), grain desserts, dairy desserts, cold cereal, and candy are main sources of added sugars in the diets of children and adolescents. Studies showed positive association between SSBs and higher blood pressure and dyslipidemia. Added sugars stimulate hepatic de novo lipogenesis and increase triglyceride level. SSBs contain high amount of fructose that significantly increase de novo lipogenesis [28].

EARLY HEALTHY FOODS AND THE LONG-TERM RISK OF CANCERS

Lifestyle factors in childhood and adolescence are associated with cancer risk in adulthood. Modified lifestyle behaviors need for primary prevention of cancer. According to studies, unhealthy behaviors early in life and persist over time can enhance the risk of some cancers including premenopausal breast, ovarian, endometrial, colon and renal cancer. Unhealthy dietary pattern reduces the quality of life and increase morbidities and mortality. More researches are needed to determine the time and degree of exposures in early childhood and adolescence correlated with risk of cancer in adulthood [29].

Unhealthy dietary pattern and excess calories intake in childhood lead to obesity. Obesity increase risk of some cancers including colon, endometrial, kidney, and postmenopausal breast cancer. It has been estimated that 15–20% of cancer deaths in the USA and 5% in European Union may be correlated with excess weight (overweight and obesity) [30]. Excess adipose tissue and lipid metabolism impact on cytokine and growth factor levels including leptin, adiponectin, resistin and tumor necrosis factor alpha. Cytokines have a carcinogenic role. Abdominal obesity increase insulin level and IGF-1 activity and decrease the synthesis of IGF-binding proteins. Increased insulin and IGF-1 activity leads to reduced sex hormone binding globulin (SHBG) synthesis. So, the levels of testosterone and estradiol increase and cell proliferation and inhibition of apoptosis can occur particularly in breast and endometrial tissues [31].

The most common form of cancer in children less than 14 years is Leukemia. Finding reported that, regular intake of oranges (or natural orange juice), bananas, vegetables and bean curd can protect the risk of Leukemia. However, frequent intake of smoked meat products correlated with increased risk of childhood leukemia [32].

Studies showed that specific food types or nutrients during childhood associated with the risk of cancers. Some foods including soy, flaxseeds, grains, nuts, fruits and vegetables contain phytoestrogens. The risk of breast cancer can be decreased by consumption of foods with phytoestrogens during adolescence [33]. However, consumption of red meat, butter, French fries, and a diet with a higher glycemic index during childhood enhanced the risk of breast cancer. According to findings, consumption of eggs, vegetable fat, fiber, and vitamin E associated with decreasing the risk of breast cancer. Moderate reduction the cancer risk was reported for consumption of fruits and vegetables and milk (34). Increased fruit intake in childhood reduced cancer risk in adulthood and enhanced energy intake in childhood increased risk of cancer mortality in adulthood [35].

Some studies could not find any significant association between dairy consumption in childhood and risk of prostate and stomach cancer risk in adulthood [36, 37]. However, high dairy consumption in childhood is correlated with enhanced colorectal cancer risk [36]. Findings of case–control study showed opposite relationship between milk and dairy products intake and childhood leukemia. Fortified milk with vitamin D inhibits the clonogenic growth of normal and malignant lymphoid B cell progenitors [32]. Protein intake through dairy products may be another reason for the inverse association of dairy consumption with leukemia. Foods rich in protein contain antioxidant tripeptide glutathione. Regarding the desirable impacts of protein consumption, antioxidant tripeptide glutathione may protect the cell from ROS-mediated DNA-damage and ROS-induced regulation of gene expression. In addition, it may detoxify the potential carcinogenic compounds [38].

Children's diet with high amount of fruit and vegetable correlated with lower risk of some cancers including digestive tract, breast, and pancreas. Fruits and vegetables contain bioactive plant chemicals including phenolics and carotenoids. These phytochemicals have antioxidant properties. They prevent or reduce oxidative stress and damage of lipids, proteins, and DNA that it can lead to cancer. Bioactive food compounds including polyphenols, isothiocyanates, allyl compounds, folate, selenium, retinoids and omega 3 fatty acids modulate epigenetic processes and affect DNA methylation, tumor suppressor gene promoter methylation and post-translational modifications of histones in different cancer cells. In addition,

BFCs could influence DNA repair, oxidative stress, inflammation, cell growth, differentiation and apoptosi. More investigations need for assessment the impact of BFCs on cancer prevention [39].

EARLY CHILDHOOD NUTRITION AND THE LONG-TERM RISK OF OSTEOPOROSIS

Since significant proportion of bone mass in adulthood is accrued during childhood and adolescence and peak bone mass attainment associated with early life nutrition, study on the role of dietary components on bone mass, bone matrix formation and preservation in early life is important [40].

Childhood and adolescence are critical periods of growth and maturation. One of the important modifiable factors in the development of bone mass during childhood is healthy diet. Bone mass greatly increases in adolescence during the growth spurt. Maximum calcium retention occurs in early puberty. Bone density normally decreases during later life and failure to achieve optimal bone mass at the end of adolescence lead to bone problem in later years [41].

One meta-analysis study assesses relationship between dietary patterns, bone mineral density (BMD), and risk of fracture. The findings showed that the risk of low BMD can be decreased by prudent/healthy dietary pattern among children and adolescents. Healthy dietary pattern contain high intakes of fruits, vegetables, whole grains, legumes, nuts, fish, low-fat dairy products, and low-fat milk, and low intakes of soft drinks, sugars, refined grains or cereals, red meat, and processed meat. Western/Unhealthy" dietary patterns in childhood that contain red meat, processed meat, soft drinks, refined grains or cereals, fast food, and sweets associated with increasing the risk of fracture in adulthood [42].

It has been reported that more sugar-sweetened beverages and soft drinks consumption have adverse impacts on health including low bone mineral density and hypocalcaemia among children and adolescents.

Sugar-sweetened beverages contain fructose, caffeine and phosphoric acid that correlated with poor bone health [42].

Adequate intake of some nutrients including calcium, phosphorus, vitamin D, vitamin K, magnesium, zinc, fluoride and protein is needed for normal bone development and bone metabolism. Vitamin D is essential for gastrointestinal absorption of calcium and phosphorus, and reduce renal excretion of calcium.11 Small number of foods including fatty fish and liver contain vitamin D. Fortified food products include cow's milk are other sources of vitamin D [43].

Children and adolescents consume inadequate amount of calcium and dairy product. They should be encouraged to intake more calcium-rich products to provide their calcium needs. Some studies showed that those who consume more calcium and dairy product have stronger bones [44, 45]. However, some studies did not show any significant association [46, 47]. Cross sectional study showed that dietary calcium intake did not significantly associate with serum calcium and biochemical indicators of bone health in 9-12 years old Iranian children [48].

Study on 2850 children participating in a population-based prospective cohort study showed that intake of dairy and whole grains dietary pattern in infancy which contains whole grains, dairy and cheese, and eggs was correlated with higher BMD in childhood and had useful impacts on bone outcomes. Dairy foods provide important nutrient including calcium, magnesium, vitamin D (especially if fortified) and high-quality proteins. Whole grain products contain magnesium, iron, B vitamins, and phytochemicals, antioxidants and other bioactive compounds, which have beneficial effects on bone health [40].

Findings showed that Mediterranean diet which contain higher intake of fruits, vegetables, whole grains, legumes, and nuts, fish, olive oil, and lower intake of red meat and saturated fatty acids, positively correlated with bone health [49].

Dietary intervention in childhood is needed to increase peak bone mass and calcium reserves and reduce the bone problems in later years. Unhealthy nutrition will enhance risk of future bone fragility and fractures.

CONCLUSION

Primary prevention strategies need for decreasing the risk of non-communicable diseases that are major public health challenge. Genetic predisposition, improper dietary behavior and sedentary lifestyle are the major reasons of diseases. Proper nutrition is one of the modifiable factors that reduce disease progression and incidence. These effects start during early life and throughout childhood. Reduced intake of whole grain, fruits, and vegetables, and increased intake of red meat, butter, French fries, unhealthy fats, including the saturated fatty acids and trans fatty acids and dietary salt in early childhood lead to increase susceptibility to diseases later in life. Multisectoral preventive approach is needed to restrict the further increasing of non-communicable diseases.

REFERENCES

[1] Magarey A, Mauch C, Mallan K, Perry R, Elovaris R, Meedeniya J, et al. Child dietary and eating behavior outcomes up to 3.5 years after an early feeding intervention: The NOURISH RCT. *Obesity* (Silver Spring, Md). 2016 Jul;24(7):1537-45.

[2] Hendrie G, Sohonpal G, Lange K, Golley R. Change in the family food environment is associated with positive dietary change in children. *The international journal of behavioral nutrition and physical activity*. 2013 Jan 7;10:4.

[3] Krolner R, Rasmussen M, Brug J, Klepp KI, Wind M, Due P. Determinants of fruit and vegetable consumption among children and adolescents: a review of the literature. Part II: qualitative studies. *The international journal of behavioral nutrition and physical activity*. 2011 Oct 14;8:112.

[4] Wedde S, Haines J, Ma D, Duncan A, Darlington G. Associations between Family Meal Context and Diet Quality among Preschool-Aged Children in the Guelph Family Health Study. *Canadian journal of dietetic practice and research: a publication of Dietitians of*

Canada = Revue canadienne de la pratique et de la recherche en dietetique: une publication des Dietetistes du Canada. 2019 Sep 12:1-7.

[5] Blaine RE, Kachurak A, Davison KK, Klabunde R, Fisher JO. Food parenting and child snacking: a systematic review. *The international journal of behavioral nutrition and physical activity.* 2017 Nov 3;14(1):146.

[6] Sonneville KR, Rifas-Shiman SL, Kleinman KP, Gortmaker SL, Gillman MW, Taveras EM. Associations of obesogenic behaviors in mothers and obese children participating in a randomized trial. *Obesity* (Silver Spring, Md). 2012 Jul;20(7):1449-54.

[7] Nyberg G, Norman A, Sundblom E, Zeebari Z, Elinder LS. Effectiveness of a universal parental support programme to promote health behaviours and prevent overweight and obesity in 6-year-old children in disadvantaged areas, the Healthy School Start Study II, a cluster-randomised controlled trial. *The international journal of behavioral nutrition and physical activity.* 2016 Jan 21;13:4.

[8] Fisher JO, Serrano EL, Foster GD, Hart CN, Davey A, Bruton YP, et al. Title: efficacy of a food parenting intervention for mothers with low income to reduce preschooler's solid fat and added sugar intakes: a randomized controlled trial. *The international journal of behavioral nutrition and physical activity.* 2019 Jan 17;16(1):6.

[9] Walsh AD, Hesketh KD, Hnatiuk JA, Campbell KJ. Paternal self-efficacy for promoting children's obesity protective diets and associations with children's dietary intakes. *The international journal of behavioral nutrition and physical activity.* 2019 Jun 28;16(1):53.

[10] Yee AZ, Lwin MO, Ho SS. The influence of parental practices on child promotive and preventive food consumption behaviors: a systematic review and meta-analysis. *The international journal of behavioral nutrition and physical activity.* 2017 Apr 11;14(1):47.

[11] Xiao L, Van't Land B, van de Worp W, Stahl B, Folkerts G, Garssen J. Early-Life Nutritional Factors and Mucosal Immunity in the Development of Autoimmune Diabetes. *Frontiers in immunology.* 2017;8:1219.

[12] Dluzniak-Golaska K, Panczyk M, Szostak-Wegierek D, Szypowska A, Sinska B. Analysis of the diet quality and dietary habits of children and adolescents with type 1 diabetes. *Diabetes, metabolic syndrome and obesity: targets and therapy.* 2019;12:161-70.
[13] Geng T, Smith CE, Li C, Huang T. Childhood BMI and Adult Type 2 Diabetes, Coronary Artery Diseases, Chronic Kidney Disease, and Cardiometabolic Traits: A Mendelian Randomization Analysis. *Diabetes care.* 2018 May;41(5):1089-96.
[14] Bartz S, Freemark M. Pathogenesis and prevention of type 2 diabetes: parental determinants, breastfeeding, and early childhood nutrition. *Current diabetes reports.* 2012 Feb;12(1):82-7.
[15] Tan J, McKenzie C, Potamitis M, Thorburn AN, Mackay CR, Macia L. The role of short-chain fatty acids in health and disease. *Advances in immunology.* 2014;121:91-119.
[16] Kelly CJ, Zheng L, Campbell EL, Saeedi B, Scholz CC, Bayless AJ, et al. Crosstalk between Microbiota-Derived Short-Chain Fatty Acids and Intestinal Epithelial HIF Augments Tissue Barrier Function. *Cell host & microbe.* 2015 May 13;17(5):662-71.
[17] Marino E, Richards JL, McLeod KH, Stanley D, Yap YA, Knight J, et al. Gut microbial metabolites limit the frequency of autoimmune T cells and protect against type 1 diabetes. *Nature immunology.* 2017 May;18(5):552-62.
[18] Kerr JA, Gillespie AN, Gasser CE, Mensah FK, Burgner D, Wake M. Childhood dietary trajectories and adolescent cardiovascular phenotypes: Australian community-based longitudinal study. *Public health nutrition.* 2018 Oct;21(14):2642-53.
[19] Donin AS, Nightingale CM, Owen CG, Rudnicka AR, Cook DG, Whincup PH. Takeaway meal consumption and risk markers for coronary heart disease, type 2 diabetes and obesity in children aged 9-10 years: a cross-sectional study. *Archives of disease in childhood.* 2018 May;103(5):431-6.
[20] Guardamagna O, Abello F, Cagliero P, Lughetti L. Impact of nutrition since early life on cardiovascular prevention. *Italian journal of pediatrics.* 2012 Dec 21;38:73.

[21] Aranceta J, Perez-Rodrigo C. Recommended dietary reference intakes, nutritional goals and dietary guidelines for fat and fatty acids: a systematic review. *The British journal of nutrition.* 2012 Jun;107 Suppl 2:S8-22.

[22] Ning H, Labarthe DR, Shay CM, Daniels SR, Hou L, Van Horn L, et al. Status of cardiovascular health in US children up to 11 years of age: the National Health and Nutrition Examination Surveys 2003-2010. *Circulation Cardiovascular quality and outcomes.* 2015 Mar;8(2):164-71.

[23] Tambalis KD, Panagiotakos DB, Psarra G, Sidossis LS. Association of cardiorespiratory fitness levels with dietary habits and lifestyle factors in schoolchildren. *Applied physiology, nutrition, and metabolism = Physiologie appliquee, nutrition et metabolisme.* 2019 May;44(5):539-45.

[24] Kelishadi R, Poursafa P. A review on the genetic, environmental, and lifestyle aspects of the early-life origins of cardiovascular disease. *Current problems in pediatric and adolescent health care.* 2014 Mar;44(3):54-72.

[25] Simell O, Niinikoski H, Ronnemaa T, Raitakari OT, Lagstrom H, Laurinen M, et al. Cohort Profile: the STRIP Study (Special Turku Coronary Risk Factor Intervention Project), an Infancy-onset Dietary and Life-style Intervention Trial. *International journal of epidemiology.* 2009 Jun;38(3):650-5.

[26] Niinikoski H, Jula A, Viikari J, Ronnemaa T, Heino P, Lagstrom H, et al. Blood pressure is lower in children and adolescents with a low-saturated-fat diet since infancy: the special turku coronary risk factor intervention project. *Hypertension* (Dallas, Tex : 1979). 2009 Jun;53(6):918-24.

[27] Van Horn L, Carson JA, Appel LJ, Burke LE, Economos C, Karmally W, et al. Recommended Dietary Pattern to Achieve Adherence to the American Heart Association/American College of Cardiology (AHA/ACC) Guidelines: A Scientific Statement From the American Heart Association. *Circulation.* 2016 Nov 29;134(22):e505-e29.

[28] Malik VS, Hu FB. Sugar-Sweetened Beverages and Cardiometabolic Health: An Update of the Evidence. *Nutrients*. 2019 Aug 8;11(8).
[29] Kerr J, Anderson C, Lippman SM. Physical activity, sedentary behaviour, diet, and cancer: an update and emerging new evidence. *The Lancet Oncology*. 2017 Aug;18(8):e457-e71.
[30] Weihrauch-Bluher S, Schwarz P, Klusmann JH. Childhood obesity: increased risk for cardiometabolic disease and cancer in adulthood. *Metabolism: clinical and experimental*. 2019 Mar;92:147-52.
[31] Garris CS, Arlauckas SP, Kohler RH, Trefny MP, Garren S, Piot C, et al. Successful Anti-PD-1 Cancer Immunotherapy Requires T Cell-Dendritic Cell Crosstalk Involving the Cytokines IFN-gamma and IL-12. *Immunity*. 2018 Dec 18;49(6):1148-61 e7.
[32] Diamantaras AA, Dessypris N, Sergentanis TN, Ntouvelis E, Athanasiadou-Piperopoulou F, Baka M, et al. Nutrition in early life and risk of childhood leukemia: a case-control study in Greece. *Cancer causes & control: CCC*. 2013 Jan;24(1):117-24.
[33] Hsieh CJ, Hsu YL, Huang YF, Tsai EM. Molecular Mechanisms of Anticancer Effects of Phytoestrogens in Breast Cancer. *Current protein & peptide science*. 2018;19(3):323-32.
[34] Karavasiloglou N, Pestoni G, Wanner M, Faeh D, Rohrmann S. Healthy lifestyle is inversely associated with mortality in cancer survivors: Results from the Third National Health and Nutrition Examination Survey (NHANES III). *PloS one*. 2019;14(6):e0218048.
[35] Wang SM, Taylor PR, Fan JH, Pfeiffer RM, Gail MH, Liang H, et al. Effects of Nutrition Intervention on Total and Cancer Mortality: 25-Year Post-trial Follow-up of the 5.25-Year Linxian Nutrition Intervention Trial. *Journal of the National Cancer Institute*. 2018 Nov 1;110(11):1229-38.
[36] van der Pols JC, Bain C, Gunnell D, Smith GD, Frobisher C, Martin RM. Childhood dairy intake and adult cancer risk: 65-y follow-up of the Boyd Orr cohort. *The American journal of clinical nutrition*. 2007 Dec;86(6):1722-9.
[37] Andersson SO, Baron J, Wolk A, Lindgren C, Bergstrom R, Adami HO. Early life risk factors for prostate cancer: a population-based

case-control study in Sweden. *Cancer epidemiology, biomarkers & prevention*: a publication of the American Association for Cancer Research, cosponsored by the American Society of Preventive Oncology. 1995 Apr-May;4(3):187-92.

[38] Wu Q, Ni X. ROS-mediated DNA methylation pattern alterations in carcinogenesis. *Current drug targets*. 2015;16(1):13-9.

[39] Ong TP, Moreno FS, Ross SA. Targeting the epigenome with bioactive food components for cancer prevention. *Journal of nutrigenetics and nutrigenomics*. 2011;4(5):275-92.

[40] van den Hooven EH, Heppe DH, Kiefte-de Jong JC, Medina-Gomez C, Moll HA, Hofman A, et al. *Infant dietary patterns and bone mass in childhood: the Generation R Study*. Osteoporosis international: a journal established as result of cooperation between the European Foundation for Osteoporosis and the National Osteoporosis Foundation of the USA. 2015 May;26(5):1595-604.

[41] Forero-Bogota MA, Ojeda-Pardo ML, Garcia-Hermoso A, Correa-Bautista JE, Gonzalez-Jimenez E, Schmidt-RioValle J, et al. Body Composition, Nutritional Profile and Muscular Fitness Affect Bone Health in a Sample of Schoolchildren from Colombia: The Fuprecol Study. *Nutrients*. 2017 Feb 3;9(2).

[42] Denova-Gutierrez E, Mendez-Sanchez L, Munoz-Aguirre P, Tucker KL, Clark P. Dietary Patterns, Bone Mineral Density, and Risk of Fractures: A Systematic Review and Meta-Analysis. *Nutrients*. 2018 Dec 5;10(12).

[43] Bailey RL, Sahni S, Chocano-Bedoya P, Daly RM, Welch AA, Bischoff-Ferrari H, et al. Best Practices for Conducting Observational Research to Assess the Relation between Nutrition and Bone: An International Working Group Summary. *Advances in nutrition* (Bethesda, Md). 2019 May 1;10(3):391-409.

[44] Julian-Almarcegui C, Gomez-Cabello A, Huybrechts I, Gonzalez-Aguero A, Kaufman JM, Casajus JA, et al. Combined effects of interaction between physical activity and nutrition on bone health in children and adolescents: a systematic review. *Nutrition reviews*. 2015 Mar;73(3):127-39.

[45] Pettinato AA, Loud KJ, Bristol SK, Feldman HA, Gordon CM. Effects of nutrition, puberty, and gender on bone ultrasound measurements in adolescents and young adults. *The Journal of adolescent health: official publication of the Society for Adolescent Medicine*. 2006 Dec;39(6):828-34.

[46] Torres-Costoso A, Gracia-Marco L, Sanchez-Lopez M, Notario-Pacheco B, Arias-Palencia N, Martinez-Vizcaino V. Physical activity and bone health in schoolchildren: the mediating role of fitness and body fat. *PloS one*. 2015;10(4):e0123797.

[47] Babaroutsi E, Magkos F, Manios Y, Sidossis LS. Lifestyle factors affecting heel ultrasound in Greek females across different life stages. *Osteoporosis international: a journal established as result of cooperation between the European Foundation for Osteoporosis and the National Osteoporosis Foundation of the USA*. 2005 May;16(5):552-61.

[48] Omidvar N, Neyestani TR, Hajifaraji M, Eshraghian MR, Rezazadeh A, Armin S, et al. Calcium Intake, Major Dietary Sources and Bone Health Indicators in Iranian Primary School Children. *Iranian journal of pediatrics*. 2015 Feb;25(1):e177.

[49] Julian C, Huybrechts I, Gracia-Marco L, Gonzalez-Gil EM, Gutierrez A, Gonzalez-Gross M, et al. Mediterranean diet, diet quality, and bone mineral content in adolescents: the HELENA study. *Osteoporosis international: a journal established as result of cooperation between the European Foundation for Osteoporosis and the National Osteoporosis Foundation of the USA*. 2018 Jun;29(6):1329-40.

In: Healthy Food
Editor: Anthony E. Walton

ISBN: 978-1-53617-599-8
© 2020 Nova Science Publishers, Inc.

Chapter 3

FUNCTIONAL INGREDIENTS AND FOOD NEOPHOBIA TOWARDS HEALTHY MEAT PRODUCTS ENRICHED WITH AGROINDUSTRIAL COPRODUCTS FLOURS

Nallely Saucedo-Briviesca[1], Alfonso Totosaus[2] and M. Lourdes Pérez-Chabela[1],*

[1]Biotechnology Department,
Universidad Autonoma Metropolitana Iztapalapa,
Mexico City, México.
[2]Food Science Lab & Pilot Plant,
Tecnologico Estudios Superiores Ecatepec. Ecatepec,
Estado de Mexico, Mexico

ABSTRACT

Carrot bagasse flour and banana peel flour have the potential to be employed as functional ingredients to improve texture, color, and flavor

---

* Corresponding Author's E-mail: lpch@xanum.uam.mx.

of raw meat products, as chorizo, or cooked meat products, as sausages. Total dietary fiber content was 40.87% and 31.67% for carrot bagasse flour and banana peel flour, respectively. The natural antioxidants content as total polyphenols were high in carrot bagasse (206 vs. 42 mg catechin/100 g flour), but the antioxidant capacity was high for banana peel flour (283 vs. 191 TEAC). In the same manner, prebiotic activity was higher for carrot bagasse flour (0.63 vs. 0.10). The composition of the flour made them a good candidate for functional meat products extenders, since incorporation in 2% and 4% in a raw meat product as chorizo, and a cooked meat product as sausage, improved yield, where the fiber hydration during process retained more water, improving expressible moisture as well. In sensory acceptation, the general appearance of banana peel flour chorizo was rejected by the consumers, whereas carrot bagasse flour chorizo was more accepted. Sausage texture was more accepted by the consumers for both flours. For cooked sausages' general appearance, carrot bagasse tendency was in the "nor like or dislike", with a better acceptance for banana peel flour samples. Nonetheless, texture has a good acceptation for both flours. However, in the neophobia test, 56% of consumers tend to reject new foods. Still, carrot bagasse flour and banana peel flour can be considered as a good option to improve the nutritional profile of raw or cooked meat products with fiber and antioxidants, with no major effect of texture.

Keywords: functional ingredients, neophobia, healthy meat products, agro industrial coproducts

1. INTRODUCTION

Food habits had been changed in recent years, since consumers are in most cases awareness for their health, given the epidemic growth of non-transmissible diseases as overweight and obesity. The demand for foods with a lower amount of additives and resistance to consuming ultra-processed foods is the first step to a healthier way to choose and consume processed foods. Nonetheless, some traditional foods, like meat products, can be enrichment with the addition of ingredients that improve their nutritional value. The inclusion of fiber with a beneficial physiological activity like oligosaccharides and other sugars that lactic acid bacteria can ferment represent a good alternative to develop and increase the availability of healthy meat products.

Now a widely exploited source for these kinds of fiber ingredients is fruit peels. Fruit peels represent around 30% of the fruit, generating a considerable amount of residues (Goñi and Hervert-Hernández, 2011). Fruit peels contain active compounds like fiber and antioxidants, that can be employed as a bioactive extensor in meat products (Lopez-Vargas et al., 2013). Examples of the use of fruit peels can be the substitution of carrageenan by orange bagasse in cooked ham, improving yield and color (Rico et al., 2011). The use of orange peel in longaniza de Pascua, a semi-dry meat product resulted in the improvement of lactobacillus, due to the putative prebiotic effect of this ingredient, besides enhancing the sensory acceptation of this meat product (Sayas-Barberá et al., 2012). Also, cactus pear peel and pineapple peel had been employed to increase fiber content in cooked meat products (Díaz-Vela et al., 2013). Residues as apple marc that contains 70% of fiber and a higher amount of phenolic compounds can be employed to retard oxidative rancidity of fat in meat products (Cerda-Tapia et al., 2015).

Fruit residues to improve the nutritional value of meat products are linked to technological characteristics, like yield, texture, and color. Orange peel improved the yield and texture of cooked sausages, enhancing water retention and decreasing oxidative rancidity (Pérez-Chabela et al., 2015). Another important aspect is the consumers' attitude to new meat products with fiber. Diaz-Vela et al. (2017) evaluated the sensory properties and food neophobia of cooked sausages containing cactus pear or pineapple peel fiber, where although sausages containing fiber are well accepted with particular approval on color and texture, there is a tendency to reject fiber-containing meat products. When the offer of meat products containing fiber from residues increase, the acceptation will be higher as well.

In this view, this work is about the use of two residues, carrot bagasse, and banana peel, that can be employed as a fiber source in two types of meat products, one raw and another one cooked. The techno-functional performance of the fiber will be different since dry products contain a lower amount of water than cooked meat products, where the moisture is maintained by the hydratable compounds, like fiber from residues.

2. Material and Methods

2.1. Agro-Industrial Coproducts Flours

Carrot (*Daucus carota L.*) bagasse and Tabasco banana (*Musa sapientum*) peel were recovered from local fresh fruit establishments in Mexico City from May to August in 2018. Bagasse and peels were dried in a Weston food dehydrator model 74-1001-w (Weston, Southern Pines) at 60°C for approximately 24 h. Dried samples were ground in a grain mill and sieved consecutively in No. 100, 80, 50 and 20 sieves to obtain a regular and homogeneous powder named flour. Different collected lots were mixed to obtain a single batch. Carrot bagasse flour and banana peel flour were stored in hermetic containers until use.

Chemical composition of peels flours were determined in accordance with AOAC Official Methods (1999) for moisture (Official Method 925.10), ashes (Official Method 942.05), total protein (Official Method 984.13, conversion factor 6.25), ethereal extract (Official Method 920.39) and total dietary fiber (Official Method 991.43). Total available carbohydrates were calculated as the difference of moisture, ashes, total protein, ethereal extract, and total dietary fiber.

2.2. Polyphenols Extraction and Antioxidant Capacity

Extraction of polyphenols and related compounds were made macerating 1 g of each flour in 100 mL of a methanol: water solution (1:2, v/v) during 4 h at room temperature with magnetic stirring. Extracts were filtered with a Whatman No. 1 filter paper and total phenolic content was determined according to Singleton and Rossi (1965) procedure.

Antioxidant capacity was carried out according to the reported by Re et al. (1999). ABTS+ radical cation was generated by reacting 7 mM 2, 2'-Azino-bis (3-ethyl benzothiazoneline-6-sulfinic acid) diammonium salt (ABTS) (Sigma Aldrich, St. Louis) and 2.45 mM potassium persulfate after incubation at room temperature in the dark for 12 hr. The ABTS+

solution was diluted in distilled water until an absorbance of 0.700 ± 0.020 at 734 nm. A Trolox solution (20 µM) was prepared for a standard curve. 495 µL of ABTS+ solution was added to 5 µL of sample and the reactive mixture was stand at room temperature for 6 min and the absorbance was immediately recorded at 734 nm. Results were expressed in terms of Trolox equivalent antioxidant capacity (TEAC, µmol Trolox equivalents per 100 g dry sample).

2.3. Prebiotic Activity Score

Culture medium was formulated employing the different flours as a carbon source to evaluate their effect on growth and acidification of the different strains. Culture medium was composed by 0.5% casein peptone (w/v), 0.3% yeast extract and carbon source at 1.0%. Total soluble carbohydrates were assessed by the Dubois et al. (1956) phenol sulphuric acid method for total reducing sugars. Glucose was employed as control and the amount of added peel flour was calculated according to the total soluble carbohydrates (total reducing sugars) to add the same amount of carbon source. Strains (10 mL with 10^7 CFU/mL) were inoculated in 90 mL of the different culture mediums serological flask (100 mL) and incubated at 37°C at anaerobic conditions. Fermentations were monitored during 8 h, sampling each hour to determine the viable count of probiotic *Lactobacillus rhamnosus GG* and *E. coli* ATCC 25922 at 0 and 8 h. Prebiotic activity score was determined as the relationship describe by Huebner et al. (2007), considering the growth of each bacteria during fermentation employing carrot bagasse flour or banana peel flour as prebiotic, according to equation 1:

$$\left\{ \frac{\text{Probiotic log CFUmL}^{-1} \text{at 8 h in prebiotic} - \text{Probiotic log CFUmL}^{-1} \text{at 0 h in prebiotic}}{\text{Probiotic log CFUmL}^{-1} \text{at 8 h in glucose} - \text{Probiotic log CFUmL}^{-1} \text{at 0 h in glucose}} \right\} - \left\{ \frac{\text{Enteric log CFUmL}^{-1} \text{at 8 h in prebiotic} - \text{Enteric log CFUmL}^{-1} \text{at 0 h in prebiotic}}{\text{Enteric log CFUmL}^{-1} \text{at 8 h in glucose} - \text{Enteric log CFUmL}^{-1} \text{at 0 h in glucose}} \right\} \quad (1)$$

2.4. Raw Meat Product: Chorizo

A total of 5 chorizo batches were elaborated with ground pork (45% w/w) and pork backfat (36% w/w) mixed with sweet paprika (3.6% w/w), salt (2% w/w), hot paprika (1.5% w/w), curing salt (0.3% w/w), garlic powder (0.3% w/w), black pepper (0.3% w/w), nutmeg (0.3% w/w), ginger powder (0.2% w/w), vinegar (2.5% v/w) and white wine (3.0% v/w), plus carrot bagasse flour or banana peel flour (2% or 4% w/w). Control treatment has no carrot bagasse flour or banana peel flour. The mixture was stored at 4°C during 24 h and stuffed in a natural casing and stored in a maturation chamber at 10°C and 50 RH for 30 days.

Chorizos of the different formulations were analyzed as follow: yield, reported as the weight of the final product after 30 days of storage in relation to original weight; pH employing a penetration electrode in a Hanna HI 99163 potentiometer. Aw was measured in an Aqualab 4TEV water activity analyzer.

Instrumental color of the internal part of the samples was determined with a Color Flex EZ HunterLab colorimeter to obtain CIE-Lab parameters: luminosity (L*), redness (a*) and yellowness (b*). Results are the average of four readings rotating each sample by 90°. From the CIE-Lab values, the hue angle (H) and saturation index (S) were calculated as described by Little (1975), according to:

$$\text{Hue angle (H)} = \text{Tan}^{-1} \frac{b^*}{a^*} \quad (2)$$

$$\text{Saturation index (S)} = \sqrt{a^{*2}+b^{*2}} \quad (3)$$

The total color difference (ΔE) in inoculated samples, considering the control sample as reference (Cava et al., 2012), was calculated as:

$$\text{Color difference } (\Delta E) = \sqrt{(L^*_{Control}-L^*)^2+(a^*_{Control}-a^*)^2+(b^*_{Control}-b^*)^2} \quad (4)$$

Textural profile analysis was performed on 20 mm height chorizo samples. Samples were compressing axially in two consecutive cycles (50% original height) with a 40 mm diameter acrylic probe at one m/s cross-head speed, a waiting period of 5 s, in a CT3 Brookfield Texture Analyzer. From the force-time curves textural parameters were calculated as follows: hardness (force necessary to attain a given deformation, maximum force), cohesiveness (strength of the internal bonds making up the body of the product), springiness (the extent to which a product returns to its original shape when compressed) (Szczesniak, 1963, Bourne, 1978).

2.5. Cooked Meat Product: Sausage

Five batches of cooked sausages were elaborated. Lean pork (50% w/w) was ground through a 0.42-cm plate in a meat grinder and mixed with salt (2% w/w), commercial phosphate mixture (0.3% w/w) and curing salt (0.3% w/w) with half of the total ice for two min in a Hamilton Chef Prep 70610 Food Processor. Frozen lard (20% w/w) was added and emulsified for 2-3 minutes. The rest of the ice was added and emulsified for 2-3 min, adding potato starch (5% w/w) and carrot bagasse flour or banana peel flour (2% or 4% w/w) until total ingredient incorporation, maintaining the batter temperature at 12 ± 2°C. Control treatment has no carrot bagasse flour or banana peel flour. The batters were stuffed into 20-mm diameter cellulose casing and cooked in a water bath until reaching an internal temperature of 70 ± 2°C (about 15 min), then cooled in an ice bath, vacuum-packed, and stored at 4°C until subsequent analysis.

The cooking yield was reported as the weight of the final cooked product 24 after cook to original raw meat batter weight. Expressible moisture was determined by adapting the methodology reported by Jauregui et al. (1981). Three pieces of Whatman #4 filter paper were weighted, folded into a thimble shape with 2 ± 0.3 g of ground meat batter sample and centrifuged at 3,000 × g for 20 min at 4°C. Expressible moisture was reported as the percentage of weight loss from the original weight of the sample. Recooking stability was determined by adapting Haq et al. (1972) methodology. Cooked sausages (ca. 30 g) were heated in 100 mL water at 70°C for 30 min; cooking stability was reported as the percentage of the weight difference between the samples before and after recooking.

Instrumental color of the internal part of the cooked sausages samples was determined as previously, calculating, in the same manner, the hue angle (H), saturation index (S) and color difference (ΔE).

Textural profile analysis was performed as well on 20 mm height cooking sausages samples, as previously described, to report hardness, cohesiveness, and springiness.

2.6. Food Neophobia

A total of 200 people (77♀/123♂, aged 30-50 years) answer a query about food neophobia. They were informed about the development of new meat products employing agro-industrial co-products, as carrot bagasse or banana peel, are a source of fiber, to formulate functional meat products (food that contains a nutrient with beneficial health effects). The dietary fiber, vegetable or fruit substance, improve health by promoting the growth of beneficial bacteria, in addition to helping cholesterol and blood glucose reduction. Carrot bagasse/banana peels are an important source of dietary fiber and antioxidants. There was no previous tasting session, but they were asked to about their willingness to try the chorizo and/or cooked sausage with carrot bagasse or banana peel, in an attempt to capture the real willingness of people to taste new meat products. The food Neophobia Scale (Pliner and Hobden, 1992) consisted of 10 statements about eating habits (1. I am constantly sampling new and different foods, reverse scale; 2. I don't trust new foods; 3. If I don't know what is in a food, I won't try it; 4. I like foods from different countries, reverse scale; 5. Ethnic food looks too weird to eat: 6. At dinner parties, I will try new food, reverse scale; 7. I am afraid to eat things I have never had before; 8. I am very particular about the foods I will eat; 9. I will eat almost anything, reverse scale; and, 10. I like to try new ethnic restaurants, reverse scale), with five graded responses from "completely disagree" (5 points) to "completely agree" (1 point). Half of the statements were written in reverse relationship to food neophobia, considering responses in the reverse value ("completely disagree" 1 point, "completely agree" 5 points). The food neophobia score was calculated as the sum of the responses with a theoretical range from 10 (neophilic) to 50 (neophobic).

2.7. Sensory Acceptation

From university students and staff personal, 50 people (23♀/27♂, aged 25-50 years) were recruited for the acceptation test of both raw and cooked meat products, chorizo and sausage. Samples of control, banana and carrot were simultaneously presented in a single session and were asked to employ 5 points balanced structured graphical hedonic scale marked with a far-left anchor of 'extremely unacceptable' and far-right anchor 'extremely acceptable', and "neither like nor dislike" in the middle. Samples of approximately 20-25 g of the different sausages formulated with carrot bagasse flour and banana peel flour and control were presented identifying samples with 3-digit random numbers. Panelists were informed that products contain a natural fiber source, carrot bagasse flour or banana peel flour, and asked to rate, according to their appreciation, a position along the scale to match their perception for general appearance, flavor, texture and fatty sensation. Ratings were converted to a numerical score (far-left anchor of the scale = 5, far-right anchor of the scale = 1) (Lim, 2011). As criteria of acceptability, the mean score was ≥ 3, corresponding to the right side of the line scale, was adopted.

2.8. Experimental Design and Data Analysis

The effect of the different percentage of agro-industrial co-products as a fiber source in raw or cooked meat products was evaluated with the proposed model:

$$y_{ij} = \mu + \alpha_I + \beta_j + \epsilon \tag{5}$$

where y_{ij} represents the meat products properties at the i-th level of flour (2 or 4%) of the j-the agro-industrial coproduct type (carrot bagasse or banana peel); μ is the overall mean; α is the main effect of fat type; and ϵ is the residual error assumed to be normally distributed with zero mean and variance $\sigma 2$ (Der and Everitt, 2001). Results were analyzed with the PROC ANOVA procedure in SAS Software v 8.0 (SAS System, Cary, NC, USA), determining as well significantly differences between means by the Duncan means test.

3. RESULTS AND DISCUSSION

Table 1 shows the composition, fiber content, polyphenols, TEAC and prebiotic index of carrot bagasse flour and banana peel flour. Moisture content between both flours was similar, this due to a similar solid structures' composition, like carbohydrates and fiber that retain water. Ashes were higher in banana peel flour, as well as protein content. Also, the ethereal extract was higher in banana peel flour. Carrot bagasse flour presented higher content of carbohydrates and both total and dietetic fiber. In the same manner, polyphenols content was 5-fold higher for carrot bagasse flour, although the antioxidant capacity was higher for the ethanolic extract of banana peel flour. The prebiotic score was three times higher for carrot bagasse.

Some differences in chemical composition could be expected since fruits composition depends on culture, season, and/or weather, with a variation in macro and micronutrients (Emaga et al., 2007). Nonetheless, the fiber content in agro-industrial coproducts had a positive influence on lipids and glucose metabolism, promoting a reduction in serum triglycerides and cholesterol, besides enhance intestinal microbiota (Macagnan et al., 2015). This is one of the main beneficial effects of these ingredients. Polyphenol content is higher in fruit and vegetable peel (Faller and Fialho, 2010), and the incorporation of these compounds in meat products inhibit lipid oxidation, maintaining the quality of the meat products during storage (Hih and Daigle, 2003). But the antioxidant activity not only depends on polyphenols content, this is, antioxidant capacity is also attributed to ascorbates, carotenoids and other compounds that present a synergistic effect to total antioxidant capacity (Babbar et al., 2011). Banana peel contains other bioactive compounds as catecholamine and anthocyanins (González-Montelongo et al., 2010).

The higher polyphenol content and antioxidant capacity of the carrot bagasse flour and banana peel flour gave an added-value to these coproducts since they are generally discarded. Since lipid oxidation is one of the main quality deterioration cause in lipid-containing foods, like meat products, the incorporation of a cheap ingredient with antioxidant properties is a good alternative to enhance quality and nutritional characteristics of meat products.

In the same manner, prebiotic carbohydrates are metabolized by specific lactic acid bacteria in the gastrointestinal tract, and hence, improve the number of these beneficial bacteria. The effectiveness of a prebiotic depends on its capacity to be selectively fermented (Huebner et al., 2007). The prebiotic potential of the coproducts flours was more than acceptable. Main carbohydrates that act like prebiotics are fructooligosaccharides, present in fruit and vegetable tissues (Wang, 2009). Prebiotics improve the gastrointestinal and immunological systems (Al-Sheraji et al., 2013).

Table 1. Agro-industrial coproducts flour from carrot bagasse and banana peel composition, antioxidant capacity and prebiotic activity score

	Carrot bagasse	Banana peel
Moisture	19.11 ± 0.81	20.14 ± 0.62
Ashes	5.15 ± 0.02	12.35 ± 0.14
Protein	1.51 ± 0.44	2.06 ± 0.21
Ethereal extract	1.05 ± 0.08	5.44 ± 0.35
Carbohydrates	32.33 ±	28.35
Fiber	40.87 ± 0.70	31.67 ± 0.69
Total polyphenols (mg Cathequin/100 g)	206.02 ± 1.13	42.08 ± 0.35
TEAC (µmol Trolox/g)	191.83 ± 0.13	283.10 ± 1.06
Prebiotic activity score	0.36 ± 0.02	0.10 ± 0.01

The effect of different percent of both flours affected the chorizo properties. First, the yield increased significantly ($P < 0.05$) when 4% of banana peel flour was added, with the lower yield in control no-added flour samples. Nonetheless, water activity was not significantly ($P > 0.05$) affected by the different treatments. The pH was significantly ($P < 0.05$) higher in carrot bagasse flour containing samples, probably to the higher prebiotic score that allowed a better growth of native lactic acid bacteria (Table 2).

For chorizo color, banana peel flour incorporation decreased significantly ($P < 0.05$) the luminosity of the samples. Hue resulted significantly ($P < 0.05$) higher for banana peel flour samples as well, but with the significantly ($P < 0.05$) lower saturation index. In the same way, banana peel flour samples obtained significantly ($P < 0.05$) higher color difference values, both above 6 (Table 2).

Chorizo texture was affected as well by the agro-industrial coproducts flours. Incorporation of these flours resulted in a harder texture, where carrot bagasse flour containing samples resulted in significantly ($P < 0.05$) higher hardness values. Carrot bagasse containing samples resulted with the significantly ($P < 0.05$) lower cohesiveness values, whereas banana peel flour samples resulted cohesively. The incorporation of flours increased chorizo springiness where the banana peel flour containing samples presented significantly ($P < 0.05$) higher values (Table 2).

Raw matured meat products have a pH between 4.9 to 5.4, acting as a protection against microbial contamination, allowing only the growth of beneficial microorganisms like lactic acid bacteria, yeast, and molds, to develop the characteristic aroma and flavor of chorizo (Feiner, 2006). In the same manner, a decrease in water activity during chorizo maturation is due to salt addition and the moisture loss during storage (Lücke, 1994).

The incorporation of different types and proportion of agro-industrial coproducts flour, like carrot bagasse and banana peel, affected the physicochemical and textural properties of the raw meat product. The hygroscopic capacity of the added flours retained moisture, reflected in a lower weight lost during storage, but maintained as well a low water activity, probably with no effect on native microflora. Besides, the prebiotic capacity of the flours improved lactic acid bacteria proliferation, reducing chorizo pH. Natural pigments in flours affected the instrumental color. Banana peel flour increased chorizo color tone, to a redder and dark coloration, besides a lower vividness of the red coloration. Carotenoids in carrot remain even after dehydration (Meléndez-Martínez et al., 2001), and hence samples with carrot bagasse flour resulted in a bright coloration with higher tonality and more clear color. In the same manner, fiber incorporated increased chorizo hardness, but since there was no more water addition, changes in texture could be due to hygroscopic nature that binds together the whole structure.

For cooked sausages, the incorporation of banana peel flour increased significantly ($P < 0.05$) the yield, with no difference between carrot bagasse containing samples and control. For expressible moisture, when 4% of flour was employed the released water was significantly ($P < 0.05$) lower, with a higher amount in control samples. In the same manner, 4% of flours resulted in significantly ($P < 0.05$) higher values for recooking stability, with the lower ones found in control samples (Table 3).

For instrumental color, carrot bagasse flour or banana peel flour significantly ($P < 0.05$) decreased sausages luminosity since control sausages with no flour were the lighter. Hue angle values were significantly ($P < 0.05$) higher when 4% of carrot bagasse flour or banana peel flour was employed, with lower values observed in control samples. Saturation index was significantly ($P < 0.05$) higher when carrot bagasse was employed at 4% and 2%, with lower values observed in control samples. The color difference was significantly ($P < 0.05$) higher when banana peel flour was employed (values above 6) (Table 3).

For texture profile analysis the sausages with banana peel flour obtained significantly ($P < 0.05$) higher values, whereas control samples resulted in the softer texture. Sausages cohesiveness was significantly ($P < 0.05$) higher for carrot bagasse flour containing samples, and control sausages with the lower values. The same tendency was observed in springiness, with significantly ($P < 0.05$) higher values in carrot bagasse flour samples (Table 3).

Table 2. Chorizo quality parameters formulated with carrot bagasse flour or banana peel flour

Parameter	Control	Carrot bagasse flour 2%	Carrot bagasse flour 4%	Banana peel flour 2%	Banana peel flour 4%
Yield (%)	76.36 ± 0.43 c	78.20 ± 0.21 b	78.76 ± 0.63 b	77.47 ± 2.16 b	80.52 ± 1.84 a
Aw	0.925 ± 0.09 a	0.917 ± 0.07 a	0.914 ± 0.08 a	0.914 ± 0.09 a	0.923 ± 0.08 a
pH	5.09 ± 0.01 b	5.02 ± 0.02 a	4.94 ± 0.01 a	4.90 ± 0.02 c	4.88 ± 0.01 c
Luminosity	34.40 ± 0.16 a	34.74 ± 0.38 a	34.73 ± 0.34 a	33.25 ± 0.63 b	34.06 ± 0.21 b
Hue angle (H)	0.744 ± 0.036 c	0.753 ± 0.076 b	0.769 ± 0.032 b	0.795 ± 0.040 a	0.787 ± 0.035 a
Saturation index (S)	34.41 ± 0.43 b	36.78 ± 0.61 a	35.26 ± 0.29 a	24.83 ± 0.15 c	27.75 ± 0.33 c
Color difference (ΔE)	–	2.41 ± 0.34 c	1.20 ± 0.23 d	10.21 ± 0.61 a	6.81 ± 0.22 b
Hardness (N)	4.29 ± 0.41 c	8.03 ± 2.73 a	7.71 ± 1.21 a	6.00 ± 1.04 b	5.93 ± 0.38 b
Cohesiveness	0.221 ± 0.030 b	0.197 ± 0.081 c	0.199 ± 0.071 c	0.227 ± 0.074 a	0.238 ± 0.081 a
Springiness	0.683 ± 0.051 c	0.737 ± 0.067 b	0.792 ± 0.057 b	0.881 ± 0.045 a	0.808 ± 0.077 a

Table 3. Cooked sausage quality parameters formulated with carrot bagasse flour or banana peel flour

Parameter	Control	Carrot bagasse flour 2%	Carrot bagasse flour 4%	Banana peel flour 2%	Banana peel flour 4%
Yield (%)	95.93 ± 2.21 b	95.73 ± 4.31 b	95.34 ± 0.23 b	97.53 ± 1.88 a	98.16 ± 1.08 a
Expressible moisture (%)	4.53 ± 0.31 a	4.40 ± 0.25 b	3.28 ± 0.19 c	3.98 ± 0.57 b	3.82 ± 0.45 c
Recooking stability (%)	93.32 ± 2.05 c	95.47 ± 3.01 b	96.72 ± 2.87 a	95.60 ± 2.54 b	97.18 ± 2.14 a
Luminosity	73.83 ± 1.17 a	72.18 ± 0.42 b	72.23 ± 0.57 b	65.51 ± 0.45 c	58.98 ± 0.37 d
Hue angle (H)	1.270 ± 0.11 c	1.273 ± 0.15 b	1.313 ± 0.78 a	1.286 0.17 b	1.327 0.15 a
Saturation index (S)	10.40 ± 0.78 d	13.93 ± 0.90 b	16.29 ± 0.89 a	13.76 ± 0.37 b	16.95 ± 0.36 a
Color difference (ΔE)	–	3.89 ± 1.09 d	6.12 ± 0.81 c	8.56 ± 1.08 b	15.07 ± 0.90 a
Hardness (N)	40.78 ± 2.75 d	43.45 ± 1.98 c	42.34 ± 2.05 c	45.71 ± 2.91 b	51.29 ± 3.01 a
Cohesiveness	0.611 ± 0.012 c	0.627 ± 0.015 a	0.627 ± 0.054 a	0.621 ± 0.024 b	0.644 ± 0.025 b
Springiness	0.816 ± 0.031 c	0.913 ± 0.017 a	0.917 ± 0.023 a	0.865 ± 0.034 b	0.848 ± 0.025 b

Fiber addition to cooked sausages increased cooking yield, reduce expressible moisture and enhance recooking stability since the flours' components were hydrated and heated, increasing the retention of water (Chang and Carpenter, 1997). In the same manner, fibers in agro-industrial flours do not change sausages pH (Grigelmo-Miguel and Martin-Belloso, 1999), since despite their prebiotic effect, at anaerobic vacuum packaging condition the growth of lactic acid bacteria is low. The incorporation of the carrot bagasse flour or banana peel flour changed the color of cooked sausages due to the presence of compounds as carotene, lutein, and lycopene, that increase the hue or tone in the final product (Leja et al., 2013). Since fiber addition improves water retention, textural properties reflect this increasing hardness forming a strong but less cohesive matrix (García et al., 2006). This change in the structure is important to consumers in the decision to buy (Resurreccion, 2004).

Figure 1 shows the results for food neophobia with an average value of 30-58. Most of the consumers were above the mean (56%), with a neophobic attitude to sausages with fibers. Below the average (44%) were neophilic persons. Healthier meat products present a goodwill purchase but with a negative impact on the expect flavor (Shan et al., 2017). The degree of neophobia in making a decision between both sausages with added fiber as a healthier food provided to respondents during the poll probably increased the acceptance for cooked sausages containing agro-industrial coproducts fibers. Food neophobia was influenced by factors such as age, socioeconomic status, marital status and gender (Pelchat and Pliner, 1995). Acceptance of functional foods depends on socio-demographic, cognitive and attitudinal factors, simultaneously if the food tastes better or worse than its conventional counterpart (Martins et al., 1997) since providing nutritional information for a new product enhance consumers' the response towards acceptance (Villegas et al., 2008).

The sensory acceptance of raw and cooked meat products is described in Figure 2. For chorizo, texture scores were above the acceptance criteria values (> 3), with similar values among the control and the samples containing carrot bagasse flour or banana peel flour. Flavor presented as well acceptable values, but relatively a higher acceptance was observed in control samples. Nonetheless, for fat sensation, the scores for chorizo without added fiber were the lower ones. However, the general acceptance of the samples obtained the lower scores, even below the criteria of acceptability when banana peel flour was added to the formulation.

For cooked sausages, samples formulated with banana peel flour obtained lower scores for flavor as compared to carrot bagasse or control no-added fiber samples. The texture of sausages was quite similar, but still lower scores were for banana peel flour. For fatty sensation, the scores obtained were very close to control, with a relatively higher value for carrot bagasse. Nonetheless, the general appearance was lower than sausages with no added fiber for both samples (Figure 2).

Prebiotic ingredients employed in foods represent a nutritional and physiological advantage since often contribute to organoleptic quality (Arihara and Ohata, 2011). Also, prebiotic ingredients have been employed to replace fat, improving texture and product stability (Jiménez-Colmenero et al., 2001).

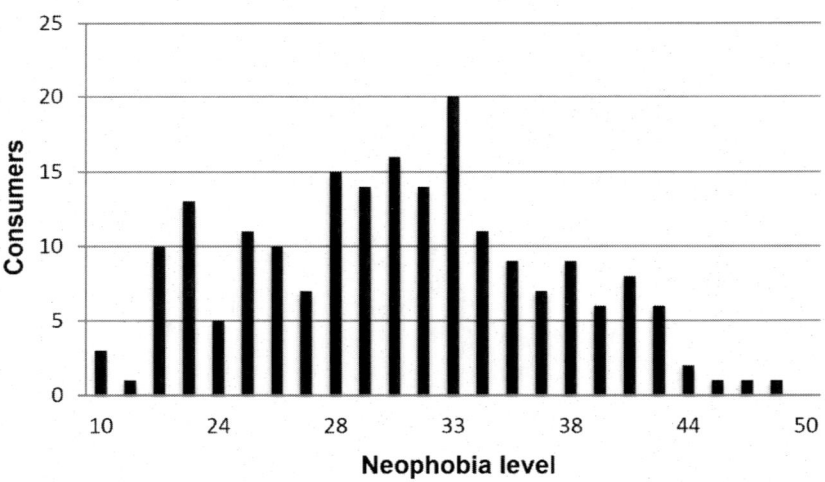

Figure 1. Food neophobia level.

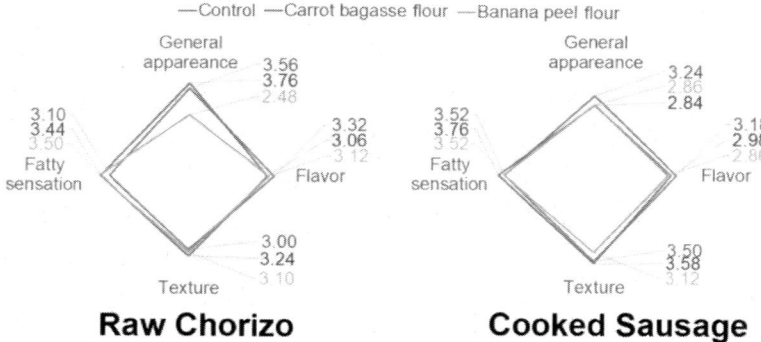

Figure 2. Sensory acceptance of raw meat product, chorizo, and a cooked meat product, sausage, formulated with 4% of carrot bagasse flour or banana peel flour.

CONCLUSION

Agro-industrial coproducts are now a tangible source for bioactive compounds looking for functional ingredients, fiber, antioxidants, and prebiotics, to enhance the nutritional profile of meat products. Meat products are related to high-fat content but with no fiber in the formulation. The use of carrot bagasse flour or banana peel flour in raw meat products enhanced the quality during storage, with no detrimental effect on color on textural characteristics. The same pattern was observed in the cooked meat product, but with a marked effect on water absorption by the fiber, resulting in a more ductile texture. The most important part of this research is the expectation of consumers toward a healthy meat product, where the fiber and their benefits could be the key to improve the acceptation of raw or cooked meat products with agro-industrial coproducts flours.

REFERENCES

Al-Sheraji, S. H., Ismail, A., Manap, M. Y., Mustafa, S., Yusof, R. M. & Hassan, F. A. (2013). Prebiotics as functional foods: a review. *Journal of Functional Foods*, 5, 1542-1553.

AOAC. (1999). *Official Methods of Analysis of AOAC International. 16th edition*. AOAC International, Gaithersburg, USA.

Arihara, K. & Ohata, M. (2011). Use of probiotics and prebiotics in meat products. In: *Processed Meats*, J.P. Kerry & J.F. Kerry (eds.). Woodhead Publishing Ltd., Cambridge, pp. 403-417.

Babbar, N., Oberoi, H. S., Uppal, D. S. & Patil, R. T. (2011). Total phenolic content and antioxidant capacity of extracts obtained from six important fruit residues. *Food Research International*, *44*, 391-396.

Bourne, M. C. (1978). Texture Profile Analysis. *Food Technology*, *32* (7), 62-66, 72.

Cava, R., Landero, L., Cantero, V. & Ramírez, M. R. (2012). Assessment of different dietary fibers (tomato fiber, beet root fiber, and inulin) for the manufacture of chopped cooked chicken products. *Journal of Food Science*, *77*, C346-C352.

Cerda-Tapia, A., de Lourdes Pérez-Chabela, M., Pérez-Álvarez, J. Á., Fernández-López, J. & Viuda-Martos, M. (2015). Valorization of pomace powder obtained from native Mexican apple (*Malus domestica* var. *rayada*): chemical, techno-functional and antioxidant properties. *Plant foods for human nutrition*, *70*, 310-316.

Chang, H. C. & Carpenter, J. A. (1997). Optimizing quality of frankfurters containing oat bran and added water. *Journal of Food Science*, *62*, 194-197.

Der, G. & Everitt, B. S. (2001). *A Handbook of Statistical Analyses using SAS*, 3rd edition. Chapman & Hall/CRC, London, pp. 99-109.

Díaz-Vela, J., Totosaus, A., Cruz-Guerrero, A. E. & Pérez-Chabela, M. (2013). *In vitro* evaluation of the fermentation of added-value agroindustrial by-products: cactus pear (*Opuntia ficus-indica* L.) peel and pineapple (*Ananas comosus*) peel as functional ingredients. *International Journal of Food Science and Technology*, *48*, 1460-1467.

Díaz-Vela, J., Totosaus, A., Escalona-Buendía, H. B. & Pérez-Chabela, M. L. (2017). Influence of the fiber from agro-industrial co-products as functional food ingredient on the acceptance, neophobia and sensory characteristics of cooked sausages. *Journal of Food Science and Technology*, *54*, 379-385.

Dubois, M., Gilles, K. A., Hamilton, J. K., Rebers, P. T. & Smith, F. (1956). Colorimetric method for determination of sugars and related substances. *Analytical chemistry*, *28*, 350-356.

Emaga, T. H., Andrianaivo, R. H., Wathelet, B., Tchango, J. T. & Paquot, M. (2007). Effects of the stage of maturation and varieties on the chemical composition of banana and plantain peels. *Food Chemistry*, *103*, 590-600.

Faller, A. L. K. & Fialho, E. F. N. U. (2010). Polyphenol content and antioxidant capacity in organic and conventional plant foods. *Journal of Food Composition and Analysis*, *23*, 561-568.

Feiner, G. (2006). *Meat Products Handbook*. Woodhead Publishing Ltd., Cambridge.

García, M. L., Cáceres, E. & Dolores Selgas, M. (2006). Effect of inulin on the textural and sensory properties of mortadella, a Spanish cooked meat product. *International Journal of Food Science and Technology*, *41*, 1207-1215.

Goñi, I. & Hervert-Hernández, D. (2011). By-products from plant foods are sources of dietary fibre and antioxidants. In: *Phytochemicals. Bioactivities and impact on health.*, *I*. Rasooli (ed.). InTech Open, London, pp. 95-116.

González-Montelongo, R., Lobo, M. G. & González, M. (2010). Antioxidant activity in banana peel extracts: testing extraction conditions and related bioactive compounds. *Food Chemistry*, *119*, 1030-1039.

Grigelmo-Miguel, N. & Martín-Belloso, O. (1999). Comparison of dietary fibre from by-products of processing fruits and greens and from cereals. *LWT-Food Science and Technology*, *32*, 503-508.

Haq, A., Webb, N. B., Whitfield, J. K. & Morrison, G. S. (1972). Development of a prototype sausage emulsion preparation system. *Journal of Food Science*, *37*, 480-484.

Hih, F. S. & Daigle, K. W. (2003). Antioxidant properties of milled-rice co-products and their effects on lipid oxidation in ground beef. *Journal of Food Science*, *68*, 2672-2675.

Huebner, J., Wehling, R. L. & Hutkins, R. W. (2007). Functional activity of commercial prebiotics. *International Dairy Journal*, *17*, 770-775.

Jauregui, C. A., Regenstein, J. M. & Baker, R. C. (1981). A simple centrifugal method for measuring expressible moisture. *Journal of Food Science*, *40*, 1271-1273.

Jiménez-Colmenero, F., Carballo, J. & Cofrades, S. (2001). Healthier meat and meat products: their role as functional foods. *Meat Science*, *59*, 5-13.

Leja, M., Kamińska, I., Kramer, M., Maksylewicz-Kaul, A., Kammerer, D., Carle, R. & Baranski, R. (2013). The content of phenolic compounds and radical scavenging activity varies with carrot origin and root color. *Plant Foods for Human Nutrition*, *68*, 163-170.

Lim, J. (2011). Hedonic scaling: A review of methods and theory. *Food Quality and Preference*, *22*, 733-747.

Little, A. C. (1975). Off on a tangent. *Journal of Food Science*, *40*, 410-411.

López-Vargas, J. H., Restrepo, D. A. & Isaza, Y. L. (2013). Oxidación lipídica y antioxidantes naturales en derivados cárnicos. *Journal of Engineering and Technology*, *2*, 50-66.

Lücke, F. K. (1994). Fermented meat products. *Food Research International*, *27*, 299-307.

Macagnan, F. T., dos Santos, L. R., Roberto, B. S., de Moura, F. A., Bizzani, M. & da Silva, L. P. (2015). Biological properties of apple pomace, orange bagasse and passion fruit peel as alternative sources of dietary fibre. *Bioactive Carbohydrates and Dietary Fibre*, *6*, 1-6.

Martins, Y., Pelchat, M. L. & Pliner, P. (1997). Try it; it's good and it's good for you: Effects of taste and nutrition information on willingness to try novel foods. *Appetite*, *28*, 89-102.

Meléndez-Martínez, A. J., Vicario, I. M. & Heredia, F. J. (2001). Estabilidad de los pigmentos carotenoides en los alimentos. *Archivos Latinoamericanos de Nutrición*, *54*, 209-215. [Stability of carotenoids pigments in foods. *Latin-American Archives of Nutrition*, *54*, 209-215].

Pelchat, M. L. & Pliner, P. (1995). Try it. You'll like it. Effects of information on willingness to try novel foods. *Appetite*, *24*, 153-165.

Pérez-Chabela, M. L., Chaparro-Hernández, J. & y Totosaus, A. (2015). Dietary fiber from agroindustrial by-products: Orange peel flour as functional ingredient in meat products. In *Dietary Fiber: Production Challenges, Food Sources and Health Benefits* (M.E. Clemens, ed.), pp. 145–157, Nova Science Publishers, Hauppauge.

Pliner, P. & Hobden, K. (1992). Development of a scale to measure the trait of food neophobia in humans. *Appetite*, *19*, 105-120.

Re, R., Pellegrini, N., Proteggente, A., Pannala, A., Yang, M. & Rice-Evans, C. (1999). Antioxidant activity applying an improved ABTS radical cation decoloration assay. *Free Radical Biology & Medicine*, *26*, 1231-1237.

Resurreccion, A. V. A. (2004). Sensory aspects of consumer choices for meat and meat products. *Meat Science*, *66*, 11-20.

Rico, J. A. A., Cruz, L. G., Goiz, J. M. J., Álvarez, Z. B. G., Ortíz, M. E. R. & Nicanor, A. B. (2011). Efecto de la utilización de bagazo de naranja como extensor funcional sobre las propiedades fisicoquímicas y texturales de jamón cocido. *Nacameh*, *5*, 27-39. [Effect of the use of orange bagasse as a functional extender on the physicochemical and textural properties of cooked ham.]

Sayas-Barberá, E., Viuda-Martos, M., Fernández-López, F., Pérez-Álvarez, J. A. & Sendra, E. (2012). Combined use of a probiotic culture and citrus fiber in a traditional sausage 'Longaniza de Pascua'. *Food control*, *27*, 343-350.

Shan, L. C., De Brún, A., Henchion, M., Li, C., Murrin, C., Wall, P. G. & y Monahan, F. J. (2017). Consumer evaluations of processed meat products reformulated to be healthier. A conjoint analysis study. *Meat Science*, *131*, 82-89.

Singleton, V. L. & Rossi, J. A. (1965). Colorimetry of total phenolics with phosphomolybdic-phosphotungstic acid reagents. *American Journal of Enology and Viticulture*, *16*, 144-158.

Szczesniak, A. S. (1963). Classification of Textural Characteristics. *Journal of Food Science*, *28*, 385-389.

Villegas, B., Carbonell, I. & Costell, E. (2008) Effect of product information and consumer attitudes on responses to milk and soybean vanilla beverages. *Journal of the Science of Food and Agriculture, 88,* 2426-2434.

Wang, Y. (2009). Prebiotics: Present and future in food science and technology. *Food Research International, 42,* 8-12.

In: Healthy Food
Editor: Anthony E. Walton

ISBN: 978-1-53617-599-8
© 2020 Nova Science Publishers, Inc.

Chapter 4

CANDELILLA WAX AS OIL RESTRUCTURING AGENT AS FAT REPLACER TO FORMULATE HEALTHY COOKED SAUSAGES

Alfonso Totosaus[*], Aislinn N. Botello-Pérez, D. Aurora Hernández-Domínguez, Octavio Toledo and B. Mariel Ferrer-González
Food Science Lab and Pilot Plant,
Tecnologico Estudios Superiores, Estado de Mexico, Mexico

ABSTRACT

Fat in animal food products represents a source of undesirable compounds like saturated fats, associated to worsen cardiovascular disease and obesity. Fat is an important component in emulsified cooked meat products, like sausages. Restructuring oils have become a techno-functional alternative to incorporate polyunsaturated oils as oleogel, improving texture and oxidative stability. The effect of candelilla wax oleogel to replace pork backfat lard was determined on cooked sausages

---

[*] Corresponding Author's E-mail: atotosaus@tese.edu.mx.

textural profile analysis, moisture, color, and sensory acceptance. Sausages formulated with oleogel resulted harder and more resilient but less cohesive than samples with lard. Oleogel increased total moisture and enhance water retention. The incorporation of oleogel to replace fat resulted in a lighter coloration, with higher tone and saturation index values, and acceptable color difference (<6). As expected, oleogel samples obtained lower oxidative rancidity than lard containing samples. In sensory acceptance, no difference between oleogel sample and control with lard was detected for color, taste, fat sensation, texture or overall acceptance. These results indicate that candelilla wax oleogel can be employed as a fat replacer in emulsified cooked sausage to reduce saturated fats and increase polyunsaturated oils, improving the health profile of this kind of meat products.

Keywords: oleogel, soybean oil, candelilla wax, fat replacement, sausages

1. Introduction

Organogels are three-dimensional, self-standing, thermo-reversible, anhydrous, viscoelastic gels created by adding organogelators and/or polymeric gelators, which can self-assemble themselves into gel networks via noncovalent interactions, entrapping the liquid phase (as vegetable edible oil) after heating/cooling, resulting gel can also be called an 'oleogel'' (Öğütcüa and Yılmaza, 2014). Among various structurants that are so far explored for gelling vegetable oils, natural waxes are considered to be the most promising ones because of their excellent oil-binding properties, economical value since are capable to gel oils at lower concentrations as 0.5% (w/v), and availability since a considerable number of waxes that are approved for use in foods (Patel, 2015). Waxes are commercially available as a mixture of organic components with a high content of wax esters and other components, displaying a broad melting profile with different temperature range depending on the wax source, leading to different oil structuring ability (Davidovich-Pinhas, 2018).

The development of crystalline structure through organogelation supply the proper rheological properties such as texture formation, creaminess, softness, aeration, plasticization, mouthfeel, etc. for different food applications. Therefore, oleogels are interesting products which are currently being used in structuring edible oils for margarine and shortening-like products, emulsion-based products, and in other processed food applications, like bakery, processed meat, ice cream and dairy, confectionery and edible films (Öğütcüa and Yılmaza, 2014). Many food manufacturers are switching on other methods of replacement of saturated fat in food products to cut the risk of cardiovascular diseases due to intake saturated fatty acids (Kaushik et al., 2017). Dietary saturated fatty acids have important effects on health, since reducing the intake of saturated fatty acids will reduce the risk of coronary heart disease. High levels of total blood cholesterol and cholesterol in low-density lipoproteins rise in risk for coronary heart disease. Dietary saturated fatty acids strongly raise total cholesterol and low-density lipoproteins cholesterol levels in the blood. Evidence from epidemiological, clinical and metabolic studies convincingly shows that replacing saturated fatty acids in the diet with polyunsaturated fatty acids is an effective way to reduce the risk of coronary heart disease. Saturated fatty acids occur in the diet in different chain lengths, with lauric, myristic, palmitic and stearic as the major ones (Zock, 2006). Seed oils are rich in polyunsaturated fatty acids (PUFAs), with commonly consumed sunflower oil and soybean oil containing approximately 60% of total fatty acids as PUFA (Minihane and Lovegrove, 2006).

Processed meats have traditionally higher fat concentrations than other fresh whole-muscle retail counterparts, largely because of the range of processing methods employed and the homogeneous of the end-products produced. During the preparation of meat and processed meat products, levels of fat can range from 20-25%. Lowering fat content of processed meats to more nutritional acceptable levels, especially below 15%, in emulsified comminuted products, such as frankfurters, gives rise to a less acceptable product with a poor flavor and increase in toughness. (Kerry and Kerry, 2006). This high-fat content restricts the consumption of meat products.

Besides, although the major role of dietary fat per se in determining body weight is not strong, the total amount of fat in the diet would in the long-term increase body weight, related to coronary heart disease and a higher risk of diabetes. Dietary fat forms only part of the equation determining energy balance, where other factors play an important role in caloric overconsumption, like high energy-dense foods, and energy expenditure, like low physical activity, are major determinants of energy imbalance and weight gain (Zock, 2006). Substitution of animal fat with vegetable oil in finely-comminuted products can lead to increased hardness and chewiness, as well as lighter and less red color. Furthermore, lipid oxidation is increased when saturated fat is replaced with unsaturated fat. Replacement of fat with vegetable oil in comminuted products can also result in processing challenges such as decreased emulsion stability in finely-comminuted products (Wolfer et al., 2018). Fatty acid composition, also, to be an important determinant of the 'health' properties of any individual fat source, has a major impact on the physical properties of fat in foodstuffs (Minihane and Lovegrove, 2006).

In this view, the use of wax oleogel as fat replacer is a novel and recently exploited alternative to reduce saturated fatty acids content in emulsified comminuted meat products, like sausages, looking for a healthier alternative.

2. MATERIALS AND METHODS

2.1. Oleogel Elaboration

Candelilla wax (3%, v/v) was dispersed by heat at 90 °C with stirring agitation in Nitrioli ® soybean oil (RAGASA, Monterrey, Mexico) with pomegranate (*Punica granatum*) oleoresin (1%, v/v). The mixture was cooled down at room temperature to form an oleogel (Toro-Vazquez et al., 2007).

2.2. Cooked Sausage Elaboration

Lean pork and lard (pork backfat) were purchased in local abattoirs, removing visible fat and connective tissue. Meat (50% w/w) was ground through a 0.42-cm plate in a meat grinder and mixed with salt (2% w/w), commercial phosphate mixture (0.3% w/w) and curing salt (0.3% w/w) with half of the total ice for two min in a Hamilton Chef Prep 70610 Food Processor. Frozen lard or frozen oleogel (20% w/w) was added and emulsified for 2-3 minutes. The rest of the ice was added and emulsified for 2-3 min, maintaining the batter temperature at 12 ± 2°C. The batters were stuffed into 20-mm diameter cellulose casing and cooked in a water bath until reaching an internal temperature of 70 ± 2°C (about 15 min), then cooled in an ice bath, vacuum-packed, and stored at 4°C until subsequent analysis.

Moisture content was quantified using AOAC Official Test Method No. 950.46 (AOAC, 1999). Two g samples were placed in an aluminum capsule at a constant weight and heated in an oven at 110 °C for 12 h. The samples were then removed, and the percentage of moisture was calculated based on weight difference. Expressible moisture was determined by adapting the methodology reported by Jauregui et al. (1981). Three pieces of Whatman #4 filter paper were weighted, folded into a thimble shape with 2 ± 0.3 g of ground meat batter sample and centrifuged at 3,000 × g for 20 min at 4°C. Expressible moisture was reported as the percentage of weight loss from the original weight of the sample.

Textural profile analysis was performed on 20 mm height chorizo samples of each treatment. Samples were compressing axially in two consecutive cycles (50% original height) with a 40 mm diameter acrylic probe at a cross-head speed of one mm/s, waiting period of 5 s, in a CT3 Brookfield Texture Analyzer. From the force-time curves textural parameters were calculated as follows: hardness (force necessary to attain a given deformation, maximum force), cohesiveness (strength of the internal bonds making up the body of the product), springiness (the extent to which a product returns to its original shape when compressed) (Szczesniak, 1963, Bourne, 1978). Resilience (energy absorbed by the sample during compression and then released, during the first compression) was determined from force-deformation curves by measuring the area enclosed by the hysteresis loop, i.e., energy stored in the sample that allows the recovery to some extent of its original shape (Voisey et al., 1975).

Instrumental color of the internal part of the samples was determined on the CIE-Lab scale employing the Color Lab application for Android O.S. (Vilka Studios). Results are the average of four readings rotating each sample by 90°. From the CIE-Lab values, the hue angle (H) and saturation index (S) were calculated as described by Little (1975), according to:

$$\text{Hue angle (H)} = \text{Tan}^{-1} \frac{b^*}{a^*}$$

$$\text{Saturation index (S)} = \sqrt{a^{*2} + b^{*2}}$$

The total color difference (ΔE) in inoculated samples, considering the control sample as reference (Cava et al., 2012), was calculated as:

$$\text{Color difference } (\Delta E) = \sqrt{(L^*_{Control} - L^*)^2 + (a^*_{Control} - a^*)^2 + (b^*_{Control} - b^*)^2}$$

The oxidative rancidity was determined with 10 g of grounded sample mixed with 49 mL of distilled water at 50°C, adding one mL of sulfanilamide-HCl solution (0.5% and 20%, respectively, v/v). Subsequently, the sample was transferred to a 500 mL Erlenmeyer flask containing 48 mL of distilled water at 50°C and 2 mL of HCl solution (50% v/v), plus 2 drops of silicone-based antifoam. The content of the flask was distilled about 10-15 minutes or until obtaining 50 mL of distillate. An aliquot of 5 mL was taken and mixed with 5 mL of thiobarbituric acid solution (0.02 M in glacial acetic acid 90%). Samples were placed in boiling water for 35 minutes, cooled and absorbance was read at 538 nm (Zipser and Watts, 1962). The concentration of malonaldehyde (mg/kg of sample) was calculated extrapolating the absorbance against a 1,1,3,3-tetraethoxypropane (3×10^3 g/L) solution.

Acceptability of meat batters was rated using a 10 cm structured graphical hedonic scale marked with a far-left anchor of 'extremely unacceptable' and far-right anchor 'extremely acceptable' (Clark and

Johnson, 2002). A total of 30 participants (22♀/18♂ in an age ranged from 20 to 45) were recruited from faculty members and students. No sensory training was provided with prior evaluation sessions. Two samples (approximately 20-25 g), with 7 days of elaboration, of control sample with lard and oleogel containing sample, were presented identified with 3-digit random numbers. Panelists were informed that in one of the samples were elaborated with restructured vegetable oil instead of pork backfat, saturated animal fat. Each subject assessed one pair of samples within two days consecutive period at approximately the same time of day. Water was supplied to clean the palate between samples. Persons were asked to rate, according to their appreciation, a position anywhere along the scale to match their perception for: taste, color, texture, fat sensation, and overall acceptance, during and after tasting the different samples. Ratings were converted to a numerical score based on the distance in mm from the far-left anchor of the scale. As criteria of acceptability, the mean score was ≥5, corresponding to the mid-point of the line scale (representing "neither acceptable nor unacceptable").

2.3. Experimental Design and Data Analysis

In order to determine the effect of the different types of fat on cooked sausage properties during 30 days of storage, the proposed model was:

$$y_{ij} = \mu + \alpha_i + \beta_j + \epsilon_{ij}$$

where y_{ij} represents the sausage total moisture, expressible moisture, instrumental texture, color parameters and oxidative rancidity for the *i*th type of fat at the *j*th day of storage; α_i and β_j are the main effects of type of fat and storage time; ϵ_{ij} represents the residual error terms, assumed to be normally distributed, with zero mean and variance σ^2 (Der and Everitt, 2008). Data analysis was carried out using SAS statistical software v. 8.0 (SAS Institute, Cary) with PROC GLM procedure. Significant difference among means was determined by Duncan's means test (P = 0.05) in the same software.

Sensory acceptance results were analyzed using a paired t-test with the PROC TTEST employing PAIRED statement and alpha = 0.05 (95% confidence interval).

3. RESULTS AND DISCUSSION

Table 1 shows the results for total and expressible moisture. Total moisture was significantly (P < 0.05) higher in sausages with oleogel than in sausages with lard. The moisture content decreased significantly (P < 0.05) with storage time. In the same manner, expressible moisture, water released when a force was applied, was significantly (P < 0.05) lower for oleogel containing samples. Expressible moisture increased significantly (P < 0.05) with storage time. This is that more water was physically retained in the meat batter matrix. Since oleogel was in a certain way easily to disperse than pork lard, small fat globules were formed, increasing fat-protein interactions resulting in less fluid exudation (Barbut et al., 2016a; Alejandre et al., 2019), increasing total moisture and decreasing expressible moisture in oleogel containing samples.

For the textural profile analysis, hardness values were significantly (P < 0.05) higher for oleogel samples. Sausages became significantly (P < 0.05) harder with storage time. Cohesiveness was significantly (P < 0.05) lower in sausages with oleogel. Cohesiveness decreased significantly (P < 0.05) with storage time. This means that oleogel resulted in a tough but easy to disintegrate texture. Springiness and resilience resulted as well significantly (P < 0.05) higher in samples containing oleogel instead of pork backfat. Springiness was not significantly (P > 0.05) affected by time, remaining constant during storage. Resilience increased significantly (P < 0.05) with storage time (Table 1). The dispersion of oleogel fat globules within the protein matrix in comminuted sausages results in firmer meat product (Barbut et al., 2016b; Barbut and Marangoni, 2019).

Table 1. Moisture and textural properties of sausages formulated with lard or oleogel during storage

Total moisture (%)		
Storage time (days)	Control	Oleogel
1	40.10 ± 0.06 b, A	46.50 ± 0.42 a, A
7	38.85 ± 0.18 b, A	46.23 ± 0.15 a, A
14	41.62 ± 1.93 b, A	47.55 ± 0.04 a, A
21	23.69 ± 0.13 b, B	31.10 ± 0.54 a, B
Expressible moisture (%)		
Storage time (days)	Control	Oleogel
1	43.54 ± 0.76 a, D	34.06 ± 0.47 b, D
7	41.03 ± 3.22 a, C	35.80 ± 2.56 b, C
14	45.46 ± 1.68 a, B	35.61 ± 0.79 b, B
21	48.09 ± 6.76 a, A	33.13 ± 1.38 b, A
Hardness (N)		
Storage time (days)	Control	Oleogel
1	11.85 ± 0.05 b, C	17.20 ± 0.87 a, C
7	18.20 ± 0.19 b, B	22.95 ± 0.93 a, B
14	20.70 ± 0.43 b, B	22.05 ± 1.26 a, B
21	24.00 ± 2.30 b, A	31.15 ± 3.34 a, A
Cohesiveness		
Storage time (days)	Control	Oleogel
1	0.695 ± 0.016 a, A	0.670 ± 0.010 b, A
7	0.690 ± 0.011 a, A	0.715 ± 0.005 b, A
14	0.700 ± 0.010 a, A	0.715 ± 0.006 b, A
21	0.710 ± 0.011 a, B	0.555 ± 0.126 b, B
Springiness		
Storage time (days)	Control	Oleogel
1	0.860 ± 0.010 b, A	0.865 ± 0.016 a, A
7	0.835 ± 0.005 b, A	0.866 ± 0.027 a, A
14	0.845 ± 0.027 b, A	0.845 ± 0.016 a, A
21	0.850 ± 0.011 b, A	0.860 ± 0.010 a, A
Resilience (N s)		
Storage time (days)	Control	Oleogel
1	98.89 ± 20.12 b, C	143.40 ± 8.25 a, C
7	97.02 ± 4.04 b, C	128.71 ± 1.56 a, C
14	120.82 ± 3.82 b, B	145.42 ± 7.70 a, B
21	131.68 ± 9.84 b, A	169.30 ± 14.55 a, A

[a, b] Means with the same letter in the same row are not significantly (P>0.05) different for the type of fat. [A, B] Means with the same letter in the same column are not significantly (P > 0.05) different for storage time.

Table 2. Color parameter and oxidative rancidity of sausages formulated with lard or oleogel during storage

Luminosity (L*)		
Storage time (days)	Control	Oleogel
1	57.88 ± 0.08 b, D	57.43 ± 0.48 a, D
7	51.38 ± 0.41 b, C	64.27 ± 0.30 a, C
14	67.02 ± 0.03 b, B	72.98 ± 0.03 a, B
21	69.93 ± 0.09 b, A	69.98 ± 0.04 a, A
Color tone (H)		
Storage time (days)	Control	Oleogel
1	82.73 ± 0.30 b, B	82.09 ± 0.10 a, B
7	81.95 ± 0.06 b, B	82.64 ± 0.37 a, B
14	83.56 ± 0.48 b, A	84.24 ± 0.16 a, A
21	83.14 ± 0.17 b, A	83.65 ± 0.38 a, A
Saturation index (S)		
Storage time (days)	Control	Oleogel
1	24.93 ± 0.15 c, C	24.28 ± 1.18 b, C
7	24.58 ± 0.32 c, B	23.57 ± 0.29 b, B
14	24.47 ± 0.44 c, B	24.43 ± 0.40 b, B
21	26.27 ± 0.65 c, A	33.62 ± 0.55 b, A
Total color difference (ΔH)		
Storage time (days)	Control	Oleogel
1	–	15.16 ± 0.18 A
7	–	5.53 ± 0.44 B
14	–	3.10 ± 0.02 C
21	–	0.00 ± 0.00 C
Oxidative rancidity (mg malonaldehyde/kg)		
Storage time (days)	Control	Oleogel
1	0.118 ± 0.001 a, D	0.177 ± 0.001 c, D
7	0.217 ± 0.000 a, C	0.326 ± 0.001 c, C
14	0.399 ± 0.001 a, B	0.279 ± 0.000 c, B
21	0.462 ± 0.000 a, A	0.314 ± 0.001 c, A

[a, b] Means with the same letter in the same row are not significantly (P>0.05) different for formulation.
[A, B] Means with the same letter in the same column are not significantly (P>0.05) different for storage time.

The differences between the effects of oleogel incorporation on sausages textural properties are related to the type of wax and percent of replacement (Barbut et al., 2016a; Alejandre et al., 2019; Franco et al., 2019). Irrespectively of differences in formulation, at the employed experimental conditions the full replacement of pork backfat lard with oleogel improved sausages texture.

The replacement of pork backfat with oleogel resulted in color changes. Luminosity was significantly ($P < 0.05$) higher in oleogel containing sausages. Luminosity increased significantly ($P < 0.05$) with storage time. Color tone and saturation index were as well significantly ($P < 0.05$) higher in oleogel sausages. For both color parameters, there was a significantly ($P < 0.05$) increase during the storage period. Color difference decreased significantly ($P < 0.05$) during storage, being lower than 6 before day 7 (Table 2). Changes in color reported as luminosity (brightness/darkness), hue angle (color tone) and saturation index (thickness and vividness of the color). According to Alejandre et al. (2019), size and dispersion of oleogel in meat batter explain color differences, since more light reflectance is related to smaller fat globules (Youssef & Barbut, 2010). Oleogel incorporation resulted in lighter and less red and yellow coloration (Barbut et al., 2016b; Kouzounis et al., 2017; Wolfer et al., 2018; Alejandre et al., 2019).

As expected, oxidative rancidity was significantly ($P < 0.05$) higher in sausages with pork backfat, increasing significantly ($P < 0.05$) as well with storage time (Table 2). The substitution of pork fat by the more polyunsaturated sunflower oil, in the form of an oleogel, did not have an adverse effect on lipid oxidation (Kouzounis et al., 2017). Besides, the use of pomegranate oleoresin with antioxidants also prevents lipid oxidation during storage.

Despite the differences detected in textural profile analysis, consumers showed no preference for the evaluated sensory attributes as sausages' color, taste, texture, fat sensation, and general acceptation. Color scores were just above the acceptance criteria, but there was significantly ($P > 0.05$) no difference between samples. The taste had as well a low score, but with no significantly ($P > 0.05$) difference. This could be related to instrumental color differences, although at day 7 the ΔE was lower than 6. When samples with vegetable oils as a fat replacer in sausages presented lower sensory and instrumental color scores there are no noticeable changes in color consumers' point of view (Câmara and Pollonio, 2015). Color could influence the initial taste of both sausages. Nonetheless, since fat is related to pleasant foods, changes in fat could mean less pleasantness for consumers, although juiciness and saltiness (taste and texture attributes) are not necessarily related to fat content (Kähkönen and Tourila, 1998). In any case, texture, even if there was no significantly ($P > 0.05$) difference between the texture of both samples, a higher score was observed in oleogel containing sample (scores above 7.0 in the 'extremely acceptable' side). In any case, fat sensation and general acceptation obtained similar scores with no significantly ($P > 0.05$) difference also. Differences of the mean rating of fat replaced samples were not lower than one rating category below the control sample, with minor values for the lower 95% confidence interval for the mean difference (Clark and Johnson, 2002).

The type of oleogel, concerning the edible oil employed and the restructuring agent, different waxes or cellulose derivatives, affect the textural and color of the emulsified meat products indeed. Besides, the percent of pork backfat or lard replacement is also important. In this research, the total replacement of saturated animal fat by the soybean oil with pomegranate resin resulted in the same acceptance scores as the control sample with pork backfat. Even with the already stated differences in formulation, oleogel in emulsified meat products can result in harder and less juicy appreciation (Barbut and Wood, 2016; Wolfer et al., 2018; Barbut and Marangoni, 2019), but with no objection as judged by panelists (Kouzounis et al., 2017).

Oleogels offer different physical characteristics as compared with vegetable oils, but with no influence or modification on chemical composition, allowing their use as fat replacers in meat products to enhance the fatty acids profile, resulting in healthier products (García-Andrade et al., 2019). The replacement of animal fat in meat products is difficult to achieve since fat within meat products naturally has a significant influence on their overall sensory properties as it builds texture, mouthfeel and accounts for a critical proportion of the overall physical food matrix, although addition of another type of ingredients with health-promoting properties may be a possible marketing tool in order to enhance the sale of fortified, low-fat meat products (Kerry and Kerry, 2006).

Table 3. Paired t-test for sensory acceptability ratings of meat batters formulated with pork backfat or oleogel as fat replacer (n= 30)

Attribute	Mean score (oleogel)	Mean score (control)	Mean difference (oleogel-control)	95% CI interval for difference	
				Inferior	Superior
(a) Color	5.26	6.01	0.75•	-0.566	2.069
(b) Taste	4.66	5.27	0.61•	-0.36	1.57
(c) Texture	8.08	6.97	1.11•	-2.18	-0.03
(d) Fat sensation	7.06	7.17	0.11•	-0.77	0.87
(e) General acceptation	6.36	6.14	0.22•	-1.23	1.35

• Value for fat replaced meat batter was not significantly different (P > 0.05) from equivalent control.

CONCLUSION

Changes in lifestyle along to the availability of healthier processed foods must be the tendency of the food industry during this century. Traditional high saturated fats emulsified comminuted meat products are being replaced by new products with less fat and improved fatty acid profile. Since last century approaches employing vegetable oils had been proposed, but now the availability of novel restructured oils with higher stability before, during and after mixing and thermal process allows to obtain tailored products with, in one hand, desirable fatty acids profile, and on the other hand, improved texture with no detrimental effect on sensory attributes. The use of soybean oleogel with pomegranate oleoresin as a natural antioxidant is a good alternative to the full replacement of pork backfat in emulsified cooked sausages.

REFERENCES

Alejandre, M., Astiasarán, I., Ansorena, D., and Barbut, S. (2019). Using canola oil hydrogels and organogels to reduce saturated animal fat in meat batters. *Food Research International*, 122, 129-136.

AOAC. (1999). *Official method 950.46: Moisture in meat. Official Methods of Analysis of AOAC International*. 16th Edition, AOAC International, Washington D.C.

Barbut, S., and Marangoni, A. (2019). Organogels use in meat processing – Effects of fat/oil type and heating rate. *Meat Science*, 149, 9-13.

Barbut, S., Wood, J., and Marangoni, A. (2016a). Quality effects of using organogels in breakfast sausage. *Meat Science*, 122, 84-89.

Barbut, S., Wood, J., and Marangoni, A. (2016b). Potential use of organogels to replace animal fat in comminuted meat products. *Meat Science*, 122, 155-162.

Bourne, M. C. 1978. Texture Profile Analysis. *Food Technology*, 32 (7), 62-66, 72

Câmara, A. K. F. I., and Pollonio, M. A. R. (2015). Reducing animal fat in bologna sausage using pre-emulsified linseed oil: technological and sensory properties. *Journal of Food Quality*, 38, 201-212.

Cava, R., Landero, L., Cantero, V., and Ramírez, M. R. (2012). Assessment of different dietary fibers (tomato fiber, beet root fiber, and inulin) for the manufacture of chopped cooked chicken products. *Journal of Food Science*, 77, C346-C352.

Clark, R., and Johnson, S. (2002). Sensory acceptability of foods with added lupin (*Lupinus angustifolius*) kernel fiber using pre-set criteria. *Journal of Food Science*, 67, 356-362.

Davidovich-Pinhas, M. (2018). Oleogels. In: *Polymeric Gels,* Pal, K. and Banerjee, I. (Eds.). Woodhead Publishing Co., Cambridge, pp. 231-249.

Der, G., and Everitt, B. S. (2001). *A Handbook of Statistical Analyses using SAS*. Chapman & Hall/CRC, London, pp. 101-116.

Franco, D., Martins, A. J., López-Pedrouso, M., Purriños, L., Cerqueira, M. A., Vicente, A. A., Pastrana, L. M., Zapata, C., and Lorenzo, J. M.

(2019). Strategy towards replacing pork backfat with a linseed oleogel in frankfurter sausages and its evaluation on physicochemical, nutritional, and sensory characteristics. *Foods*, 8, 366.

García-Andrade, M., Gallegos-Infante, J. A., and González-Laredo, R. F. (2019). Organogeles como mejoradores del perfil lipídico en matrices cárnicas y lácteas. *CienciaUAT*, 14(1), 121-132. [Organogels to improve lipidic profile in meat and dairy matrices. *CienciaUAT*, 14(1), 121-132.]

Jauregui, C. A., Regenstein, J. M., and Baker, R. C. (1981). A simple centrifugal method for measuring expressible moisture, a water binding property of muscle foods. *Journal of Food Science*, 46, 1271-1273.

Kähkönen, P., and Tourila, H. (1998). Effect of reduced-fat information on expected and actual hedonic and sensory ratings of sausage. *Appetite*, 30, 13-23.

Kaushik, I., Jain, A., Grewal, R. B., Siddiqui, S., Hahlot, R., and Anju, R. (2017). Organogelation: it's food application. *MOJ Food Processing and Technology*, 4(2), 66-72.

Kerry, J. F., and Kerry, J. P. (2006). Producing low-fat meat products. In: *Improving the Fat Content of Foods*, Williams, C., and Buttriss, J. (Eds.). CRC Press, Boca Raton, pp. 336-379.

Kouzounis, D., Lazaridou, A., and Katsanidis, E. (2017). Partial replacement of animal fat by oleogels structured with monoglycerides and phytosterols in frankfurter sausages. *Meat Science*, 130, 38-46.

Little, A. C. 1975. Off on a tangent. *Journal of Food Science*, 40, 410-411.

Minihane, A. M., and Lovegrove, J. A. (2006). Health benefits of polyunsaturated fatty acids (PUFAs). In: *Improving the Fat Content of Foods*, Williams, C., and Buttriss, J. (Eds.). CRC Press, Boca Raton, pp. 107-140.

Öğütcüa, M., and Yılmaza, E. (2014). Oleogels of virgin olive oil with carnauba wax and monoglyceride as spreadable products. *Grasas y Aceites*, 65, e040.

Patel, A. R. (2015). Natural waxes as oil structurants. In *Alternative Routes to Oil Structuring*, Springer Briefs in Food, Health and Nutrition, Cham, pp. 15-27.

Szczesniak, A. S. (1963). Classification of textural characteristics. *Journal of Food Science*, 28, 385-389.

Toro-Vázquez J. F, Morales-Rueda, J. A., Dibildox-Alvarado, E., Charó-Alonso, M., Alonzo-Macías, M., and González-Chávez, M. M. (2007). Thermal and textural properties of organogels developed by candelilla wax in safflower oil. *Journal of the American Oil Chemist Society*, 84, 989-1000.

Voisey, P. W., Randall, C., and Larmond, E. (1975). Selection of an objective test of wiener texture by sensory analysis. *Canadian Institute of Food Science and Technology Journal*, 8, 23-29.

Wolfer, T. L., Acevedo, N., Prusa, J. K, Sebranek, J. G., and Tarté, R. (2018). Replacement of pork fat in frankfurter-type sausages by soybean oil oleogels structured with rice bran wax. *Meat Science*, 145, 352-362.

Zipser, M., and Watts, B. (1962). A modified 2-tiobarbituric acid (TBA) method for determination of malonaldehyde in cured meats. *Food Technology*, 17(7), 102-104.

Zock, P. L. (2006). Health problems associated with saturated and trans fatty acids intake. In: *Improving the Fat Content of Foods,* Williams, C., and Buttriss, J. (Eds.). CRC Press, Boca Raton, pp. 3-24.

BIBLIOGRAPHY

31-day food revolution: heal your body, feel great, and transform your world	
LCCN	2018030124
Type of material	Book
Personal name	Robbins, Ocean, 1973- author.
Main title	31-day food revolution: heal your body, feel great, and transform your world / Ocean Robbins; foreword by Joel Fuhrman, MD.
Edition	First edition.
Published/Produced	New York: Grand Central Life & Style, 2019.
Description	xiii, 363 pages; 22 cm
ISBN	9781538746257 (hardcover)
LC classification	TX741 .R59 2019
Variant title	Thirty one-day food revolution
Related names	Fuhrman, Joel, writer of foreword.
Contents	part One. Detoxify -- The food revolution diet plan -- Know what's right for you -- Foods to eat and foods to avoid -- Vote with your dollars -- Build healthy eating habits -- You deserve a toxin-free home -- How to make a happy,

	healthy kitchen -- part Two. Nourish -- Eat to beat cancer -- Heal your gut -- Is breakfast sabotaging your day? -- The world's best snacks -- How to love eating vegetables -- The healthiest way to add flavor -- Enjoy healthy and delicious pleasures -- Get the goods on grains and gluten -- Legumes for long life -- What about meat and dairy? -- part Three. Gather -- Bring friends and family along -- Find a healthy eating ally -- Start a healthy meal swap team -- Eat well when you eat out -- The stunning neuroscience of gratitude -- Feed our children well -- What about school lunches? -- part Four. Transform -- GMOs and the food giants -- Is organic worth the cost? -- The simple act of growing food -- Eat for a healthy world -- Make your food cruelty-free -- Stand up for healthy food for all -- Seize the day: a time for action -- part Five. Recipes for health -- Breakfasts -- Snacks and shares -- Satisfying soups and stews -- Salads and dressings -- Entrees and greens -- Treats and desserts.
Subjects	Cooking (Natural foods)
	Nutrition.
	Diet therapy.
Notes	Includes bibliographical references (pages 332-351) and index.

Baby food universe: raise adventurous eaters with a whole world of flavorful purées and toddler foods

LCCN	2017018609
Type of material	Book
Personal name	Al-Jabbouri, Kawn, author.

Main title	Baby food universe: raise adventurous eaters with a whole world of flavorful purées and toddler foods / Kawn Al-Jabbouri, with Gemma Bischoff, R.D.
Published/Produced	Beverly, MA: Fair Winds Press, 2017.
Description	192 pages: illustrations (chiefly color); 24 cm
ISBN	9781592337477 (paperback)
LC classification	RJ216 .A464 2017
Related names	Bischoff, Gemma, author.
Summary	"Learn to make healthy food for your baby and toddler while introducing new flavors and inspiring your children to be adventurous eaters! Baby Food Universe provides healthy baby food recipes that range from simple one- and two-ingredient purees to healthy and creative toddler food. When babies first start on solid foods, they are given one-ingredient purees so parents can first check for any adverse reaction or allergy. As they get older and their likes and tolerances are known, the number of ingredients and variety can increase. This book is intended to follow those stages of development so parents can use the book for the first year and beyond. Most purees will be veggie and fruit-based, with additional foods, such as healthy fats, spices, grains, and meat, being introduced as baby's palette and tolerances grow. Includes more than 100 recipes, plus tips and advice on starting your baby on solids and cultivating healthy and happy eaters for life"-- Provided by publisher.
Subjects	Infants--Nutrition.
	Baby foods.

	Quick and easy cooking.
	Cooking / Baby Food.
	Cooking / Methods / Quick & Easy.
Notes	"Includes step-by-step photos!"--Cover.
	Includes index.

Bean Sprouts kitchen: simple and creative recipes to spark kids' appetites for healthy food

LCCN	2018023069
Type of material	Book
Personal name	Seip, Shannon Payette, author.
Main title	Bean Sprouts kitchen: simple and creative recipes to spark kids' appetites for healthy food / Shannon Payette Seip and Kelly Parthen, creators of Bean Sprouts café; photos by Lynn Renee Photography, Shannon Payette Seip, and Amy Lynn Schereck.
Published/Produced	Beverly: Fair Winds Press, 2018.
Description	1 online resource.
ISBN	9781631595486 (e-book)
LC classification	TX833.5
Related names	Parthen, Kelly, author.
Subjects	Bean Sprouts (Cafe)
	Quick and easy cooking.
	Children--Nutrition.
Form/Genre	Cookbooks.
Notes	Includes index.
Additional formats	Print version: Seip, Shannon Payette, author. Bean Sprouts kitchen Beverly: Fair Winds Press, 2018 9781592338498 (DLC) 2018021707

Bean Sprouts kitchen: simple and creative recipes to spark kids' appetites for healthy food	
LCCN	2018021707
Type of material	Book
Personal name	Seip, Shannon Payette, author.
Main title	Bean Sprouts kitchen: simple and creative recipes to spark kids' appetites for healthy food / Shannon Payette Seip and Kelly Parthen, creators of Bean Sprouts café; photos by Lynn Renee Photography, Shannon Payette Seip, and Amy Lynn Schereck.
Published/Produced	Beverly: Fair Winds Press, 2018.
ISBN	9781592338498 (pbk.)
LC classification	TX833.5 .S453 2018
Related names	Parthen, Kelly, author.
Subjects	Bean Sprouts (Cafe)
	Quick and easy cooking.
	Children--Nutrition.
Form/Genre	Cookbooks.
Notes	Includes index.
Additional formats	Online version: Seip, Shannon Payette, author. Bean Sprouts kitchen Beverly: Fair Winds Press, 2018 9781631595486 (DLC) 2018023069

Breakfast	
LCCN	2013015688
Type of material	Book
Personal name	Parker, Victoria, author.
Main title	Breakfast / Vic Parker.
Published/Produced	Chicago, Illinois: Heinemann Library, [2014]
Description	32 pages: color illustrations; 23 cm.
ISBN	9781432991166 (hb)

	1432991167 (hb)
	9781432991210 (pb)
	1432991213 (pb)
LC classification	TX355 .P254 2014
Summary	"Read Breakfast to learn how to make healthy food choices during this important meal. Different photos show healthy and unhealthy breakfast options, while simple text explains why some choices are better than others. A breakfast foods quiz concludes the book."--Provided by publisher.
Contents	Why make healthy choices? -- What makes a breakfast healthy or unhealthy? -- Processed cereals -- Granola, Muesli, and Oatmeal -- Toast -- Eggs -- Breakfast sandwiches -- Pancakes and waffles -- Fruity breakfasts -- Cooked breakfasts -- Drinks -- Food quiz -- Food quiz answers -- Tips for healthy eating.
Subjects	Nutrition--Juvenile literature.
	Breakfasts--Juvenile literature.
	Health--Juvenile literature.
Notes	Includes bibliographical references (page 32) and index.
Series	Healthy food choices
	Heinemann first library

Brian McDermott's Donegal table: delicious everyday cooking.

LCCN	2018438601
Type of material	Book
Personal name	McDermott, Brian, author.
Main title	Brian McDermott's Donegal table: delicious everyday cooking.
Published/Produced	Dublin: The O'Brien Press, 2018.

Description	192 pages: color illustrations; 26 cm
ISBN	9781847179791 (hardback)
	1847179797 (hardback)
LC classification	TX717.5 .M3338 2018
Portion of title	Donegal table: delicious everyday cooking
Summary	Brian McDermott has built a national reputation as a chef on one simple belief - that tasty, healthy food based around traditional recipes and local produce is something every family can make and enjoy. As one of twelve children growing up in Burt in County Donegal, the focal point of the family was always his mother's kitchen table, and that childhood memory of the family coming together and connecting over her home cooking continues to inspire Brian as he shares his own passion for food with others. Whether it's cooking freshly-caught mussels for the fishermen at Greencastle pier or sharing his skills with others at his cookery school, Brian loves to celebrate the best of his home county's warmth and traditions. -- Amazon
Subjects	Cooking, Irish.
Notes	Includes index.

Cat care

LCCN	2016006884
Type of material	Book
Personal name	Gardeski, Christina Mia, author.
Main title	Cat care / by Christina Mia Gardeski.
Published/Produced	North Mankato, Minnesota: Capstone Press, [2017]
Description	24 pages: color illustrations; 24 x 29 cm

ISBN	9781515709572 (library binding)
LC classification	SF447 .G24925 2017
Summary	"Simple text and full-color photos introduce techniques to take care of cats"-- Provided by publisher.
Contents	Cat care -- Healthy food -- Fresh water -- Time to play -- Nail care -- Good grooming -- Clean litter -- Safe indoors -- A check-up.
Subjects	Cats--Juvenile literature.
	Cats--Health--Juvenile literature.
Notes	Includes bibliographical references and index.
	Ages 4-8.
	K to grade 3.
Series	Pebble plus. Cats, cats, cats

Citizen farmers: the biodynamic way to grow healthy food, build thriving communities, and give back to the Earth

LCCN	2013945633
Type of material	Book
Personal name	Joffe, Daron, author.
Main title	Citizen farmers: the biodynamic way to grow healthy food, build thriving communities, and give back to the Earth / by Daron "Farmer D" Joffe; with Susan Puckett; photography by Rinne Allen.
Published/Produced	New York: Stewart, Tabori & Chang, an imprint of Abrams, 2014.
Description	224 pages: color illustrations; 24 cm.
Links	Contributor biographical information http://www.loc.gov/catdir/enhancements/fy1503/2013945633-b.html
	Publisher description http://www.loc.gov/catdir/enhancements/fy1503/2013945633-d.html

ISBN	9781617691010
	1617691011
LC classification	SB453.5 .J64 2014
Related names	Puckett, Susan, 1956- author.
	Allen, Rinne, 1973- photographer.
Summary	"In this engaging call to action, Daron Joffe teaches us not only to create sustainable gardens but also to develop a more holistic, community-minded approach to how our food is grown and how to better live our lives in balance with nature. Here is an indispensable resource packed with advice on establishing a biodynamic garden, composting, soil composition and replenishment, controlling pests and disease, cooperative gardening practices, and even creating delicious meals."--Back cover.
Contents	Composting = stewardship -- Planning = vision -- Tilling = initiative -- Sowing = faith -- Growing = patience -- Healing = compassion -- Reaping = gratitude -- Sharing = generosity -- Sustaining = perseverance.
Subjects	Biodynamic agriculture.
	Organic gardening.
	Sustainable agriculture.
	Gardening.
	Sustainable living.
Notes	Includes bibliographical references (pages 218-219) and index.

Diabetes & carb counting	
LCCN	2017933632
Type of material	Book

Personal name	Shafer, Sherri, author.
Main title	Diabetes & carb counting / by Sherri Shafer, RD, CDE, Diabetes educator.
Published/Produced	Hoboken, NJ: John Wiley & Sons, Inc., [2017]
	©2017
Description	xvii, 384 pages: illustrations; 24 cm.
ISBN	9781119315643 (paperback)
	1119315646 (paperback)
LC classification	RC662 .S528 2017
Variant title	Diabetes and carb counting
	Diabetes & carb counting for dummies
Summary	Living with diabetes doesn't have to mean giving up all of your favorite foods. Carbs from healthy foods boost nutrition and supply essential fuel for your brain and body. Counting carbs is integral to managing diabetes because your carb choices, portion sizes, and meal timing directly impact blood glucose levels. Diabetes & Carb Counting For Dummies provides essential information on how to strike a balance between carb intake, exercise, and diabetes medications while making healthy food choices.-- Source other than Library of Congress.
Contents	Getting started with carb counting and diabetes management -- Carb counting: from basic to advanced -- Going beyond counting carbs -- Embracing whole health and happiness -- Sampling menus complete with carb counts -- The part of tens.
Subjects	Diabetes--Diet therapy.
	Diabetes--Nutritional aspects.
	Food--Carbohydrate content.

	Carbohydrates.
	Diabetes Mellitus, Type 1--diet therapy.
	Diabetes Mellitus, Type 2--diet therapy.
	Diabetes Mellitus--diet therapy.
	Medical / Nutrition.
	Carbohydrates.
	Diabetes--Diet therapy.
	Diabetes--Nutritional aspects.
	Food--Carbohydrate content.
Notes	Includes index.
Series	For dummies

Diet and nutrition sourcebook: basic consumer health information about dietary guidelines, servings and portions, recommended daily nutrient intakes and meal plans, vitamins and supplements, weight loss and maintenance, nutrition for different life stages and medical conditions, and healthy food choices; along with details about government nutrition support programs, a glossary of nutrition and dietary terms, and a directory of resources for more information.

LCCN	2016024893
Type of material	Book
Main title	Diet and nutrition sourcebook: basic consumer health information about dietary guidelines, servings and portions, recommended daily nutrient intakes and meal plans, vitamins and supplements, weight loss and maintenance, nutrition for different life stages and medical conditions, and healthy food choices; along with details about government nutrition support programs, a glossary of nutrition and dietary terms, and a directory of resources for more information.
Edition	Fifth edition.

Published/Produced	Detroit, MI: Omnigraphics, [2016]
Description	xvii, 625 pages: illustrations; 24 cm.
ISBN	9780780813830 (hardcover: alk. paper)
LC classification	RA784 .D534 2016
Related names	Omnigraphics, Inc., issuing body.
Summary	"Provides basic consumer health information about nutrition through the lifespan including facts about dietary guidelines and eating plans, weight control, and related medical concerns. Includes index, glossary of related terms, and other resources"-- Provided by publisher.
Subjects	Nutrition--Popular works.
	Diet--Popular works.
	Health--Popular works.
	Consumer education.
Notes	Includes bibliographical references and index.
Additional formats	Online version: Diet and nutrition sourcebook Fifth edition. Detroit, MI: Omnigraphics, [2016] 9780780814127 (DLC) 2016026051
Series	Health reference series

Diet and nutrition sourcebook: basic consumer health information about dietary guidelines, servings and portions, recommended daily nutrient intakes and meal plans, vitamins and supplements, weight loss and maintenance, nutrition for different life stages and medical conditions, and healthy food choices, along with details about government nutrition support programs, a glossary of nutrition and dietary terms, and a directory of resources for more information

LCCN	2019021861
Type of material	Book
Main title	Diet and nutrition sourcebook: basic consumer health information about dietary guidelines, servings and portions, recommended daily

	nutrient intakes and meal plans, vitamins and supplements, weight loss and maintenance, nutrition for different life stages and medical conditions, and healthy food choices, along with details about government nutrition support programs, a glossary of nutrition and dietary terms, and a directory of resources for more information / Angela Williams, managing editor.
Edition	6th edition.
Published/Produced	Detroit, MI: Omnigraphics, Inc., [2019]
Description	1 online resource.
ISBN	9780780817128 (ebook)
LC classification	RA784
Related names	Williams, Angela, 1963- editor.
	Omnigraphics, Inc., issuing body.
Summary	"Provides basic consumer health information about nutrition through the lifespan including facts about dietary guidelines and eating plans, weight control, and related medical concerns. Includes index, glossary of related terms, and other resources"-- Provided by publisher.
Subjects	Nutrition--Popular works.
	Diet--Popular works.
	Health--Popular works.
	Consumer education.
Notes	Includes bibliographical references and index.
Additional formats	Print version: Diet and nutrition sourcebook 6th edition. Detroit, MI: Omnigraphics, Inc., [2019] 9780780817111 (DLC) 2019021352

Diet and nutrition sourcebook: basic consumer health information about dietary guidelines, servings and portions, recommended daily nutrient intakes and meal plans, vitamins and supplements, weight loss and maintenance, nutrition for different life stages and medical conditions, and healthy food choices, along with details about government nutrition support programs, a glossary of nutrition and dietary terms, and a directory of resources for more information	
LCCN	2019021352
Type of material	Book
Main title	Diet and nutrition sourcebook: basic consumer health information about dietary guidelines, servings and portions, recommended daily nutrient intakes and meal plans, vitamins and supplements, weight loss and maintenance, nutrition for different life stages and medical conditions, and healthy food choices, along with details about government nutrition support programs, a glossary of nutrition and dietary terms, and a directory of resources for more information / Angela Williams, managing editor.
Edition	6th edition.
Published/Produced	Detroit, MI: Omnigraphics, Inc., [2019]
ISBN	9780780817111 (hard cover: alk. paper)
LC classification	RA784 .D534 2019
Related names	Williams, Angela, 1963- editor.
	Omnigraphics, Inc., issuing body.
Summary	"Provides basic consumer health information about nutrition through the lifespan including facts about dietary guidelines and eating plans, weight control, and related medical concerns. Includes index, glossary of related terms, and other resources"-- Provided by publisher.

Subjects	Nutrition--Popular works.
	Diet--Popular works.
	Health--Popular works.
	Consumer education.
Notes	Includes bibliographical references and index.
Additional formats	Online version: Diet and nutrition sourcebook 6th edition. Detroit, MI: Omnigraphics, Inc., [2019] 9780780817128 (DLC) 2019021861

Diet, microbiome and health

LCCN	2017936535
Type of material	Book
Main title	Diet, microbiome and health / edited by Alina Maria Holban, Alexandru Mihai Grumezescu.
Published/Produced	London, United Kingdom; San Diego, CA, United States: Academic Press, an imprint of Elsevier, [2018]
Description	xxvi, 499 pages: illustrations; 24 cm.
ISBN	9780128114407 (paperback)
	0128114401 (paperback)
LC classification	QP145 .D54 2018
Related names	Holban, Alina Maria, editor.
	Grumezescu, Alexandru Mihai, editor.
Summary	Diet, Microbiome and Health, Volume 11, in the Handbook of Food Bioengineering series, presents the most up-to-date research to help scientists, researchers and students in the field of food engineering understand the different microbial species we have in our guts, why they are important to human development, immunity and health, and how to use that understanding to further promote research to create healthy food products. In addition, the book provides studies

	that clearly demonstrate how dietary preferences and social behavior significantly impact the diversity of microbial species in the gut and their numeric values, which may balance health and disease.-- Source other than Library of Congress.
Contents	Gut microbes: the miniscule laborers in the human body / Suma Sarojini -- Role of probiotics toward the improvement of gut health with special reference to colorectal cancer / Mian K. Sharif, Sana Mahmood, Fasiha Ahsan -- Therapeutic aspects of probiotics and prebiotics / Asif Ahmad, Sumaira Khalid -- Lactic acid bacteria beverage contribution for preventive medicine and nationwide health problems in Japan / Akira Kanda, Masatoshi Hara -- Gut microbiota alterations in people with obesity and effect of probiotics treatment / Edwin E. Martínez Leo, Armonda M. Martín Ortega, Maira R. Segura Campos -- Safety of probiotics / Dorota Zielińska, Barbara Sionek, Danuta Kołożyn-Krajewska -- Flavonoids in foods and their role in healthy nutrition / Silvia Tsanova-Savova, Petko Denev, Fanny Ribarova -- The role of milk oligosaccharides in host-microbial interactions and their defensive function in the gut / Sinead T. Morrin, Jane A. Irwin, Rita M. Hickey -- Nutritional yeast biomass: characterization and application / Monika E. Jach, Anna Serefko -- Effect of diet on gut microbiota as an etiological factor in Autism Spectrum Disorder / Afaf El-Ansary, Hussain Al Dera, Rawan Aldahash -- Dietary fibers: a way to a healthy microbiome / Prerna

	Sharma, Chetna Bhandari, Sandeep Kumar, Bhoomika Sharma, Priyanka Bhadwal, Navneet Agnihotri -- Effects of the gut microbiota on Autism Spectrum Disorder / Nalan H. Noğay -- Diet, microbiome, and neuropsychiatric disorders / Gabriel A. Javitt, Daniel C. Javitt -- Gastrointestinal exposome for food functionality and safety / Yuseok Moon -- Risk from viral pathogens in seafood / Samanta S. Khora.
	Section 1: State of the Art and Applications -- Section 2: Probiotics and Prebiotics -- Section 3: Nutritional Aspects -- Section 4: Health, Disease, and Therapy -- Section 5: Function and Safety.
Subjects	Gastrointestinal system--Microbiology.
	Biotic communities.
	Gastrointestinal Microbiome.
	Gastrointestinal Tract--microbiology.
	Biotic communities.
	Gastrointestinal system--Microbiology.
Notes	Includes bibliographical references and index.
Series	Handbook of food bioengineering; volume 11

Dinner	
LCCN	2013015691
Type of material	Book
Personal name	Parker, Victoria, author.
Main title	Dinner / Vic Parker.
Published/Produced	Chicago, Illinois: Heinemann Library, [2014]
Description	32 pages: color illustrations; 23 cm.
ISBN	9781432991180 (hb)
	1432991183 (hb)

	9781432991234 (pb)
	143299123X (pb)
LC classification	TX355 .P2545 2014
Summary	"Read Dinner to learn how to make healthy food choices during this evening meal. Different photos show healthy and unhealthy dinner options, while simple text explains why some choices are better than others. A dinner foods quiz concludes the book."-- Provided by publisher.
Contents	Why make healthy choices? -- What makes a dinner healthy or unhealthy? -- Meat -- Pizza -- Pasta -- Rice -- Fish -- Burritos and hot wraps -- Grilling -- Desserts -- Drinks -- Food quiz -- Food quiz answers -- Tips for healthy eating.
Subjects	Nutrition--Juvenile literature.
	Dinners and dining--Juvenile literature.
	Health--Juvenile literature.
Notes	Includes bibliographical references (page 32) and index.
Series	Healthy food choices
	Heinemann first library

Eating traditional food: politics, identity and practices	
LCCN	2016021306
Type of material	Book
Main title	Eating traditional food: politics, identity and practices / edited by Brigitte Sébastia.
Published/Produced	London; New York: Routledge, Taylor & Francis Group/Earthscan from Routledge, 2017.
Description	xiv, 226 pages; 24 cm.
ISBN	9781138187009 (hbk)
LC classification	GT2850 .E37 2017

Related names	Sebastia, Brigitte, editor.
Contents	Eating Traditional Food: Politics, Identity and Practices / Brigitte Sébastia -- The Rediscovery of Native "Super-foods" in Mexico / Esther Katz and Elena Lazos -- Lost in Tradition: An Attempt to Go Beyond Labels, Taking Maltese Food Practices as a Primary Example / Elise Billiard -- The Protection of Traditional Local Foods through Geographical Indications in India / Delphine Marie-Vivien -- Are Buuz and Ban¿ Traditional Mongolian Foods?: Strategy of Appropriation and Identity Adjustment in Contemporary Mongolia / Sandrine Rulhmann -- "Beef is our Secret of Life": Controversial Consumption of Beef in Andhra Pradesh, India / Brigitte Sébastia -- Modernity, Traditionalism, and the Silence Protest: The Palestinian Food Narrative in Israeli Reality Television / Liora Gvion -- The Never-ending Reinvention of 'Traditional Food": Food Practices and Identity (Re)Construction among Bolivian Returnees from Argentina / Charles-Édouard de Suremain -- What Is Healthy Diet?: Some Ideas about the Construction of Healthy Food in Germany since the 19th Century / Detlef Briesen -- Are Traditional Foods and Eating Patterns Really Good for Health?: A Socio-Anthropological Inquiry into French People with Hypercholesterolaemia / Tristan Fournier -- Eating Ayurvedic Foods: Elaboration of a Repertoire of "Traditional Foods" in France / Nicolas Commune.
Subjects	Food habits--Cross-cultural studies.
	Food consumption--Cross-cultural studies.

	Nutrition--Cross-cultural studies.
	Food--Cross-cultural studies.
	Social psychology--Cross-cultural studies.
Notes	Includes bibliographical references and index.
Series	Routledge studies in food, society and environment

Expect the best	
LCCN	2016050533
Type of material	Book
Personal name	Ward, Elizabeth M., author.
Main title	Expect the best / Elizabeth M. Ward & the Academy of Nutrition and Dietetics.
Edition	Revised edition.
Published/Produced	Nashville, Tennessee: Turner Publishing, [2017]
ISBN	9781681626246 (pbk.: alk. paper)
LC classification	RG559 .W36 2017
Related names	Academy of Nutrition and Dietetics.
Contents	Prepregnancy: starting from a healthy place -- Great expectations: how eating healthy food is good for you and your -- Baby -- Myplate plans: what to eat before, during, and after pregnancy -- Your pregnancy: expect the best -- The fourth trimester: after the baby arrives -- Food safety and other concerns: before, during, and after pregnancy -- Common concerns and special situations -- Quick and delicious recipes.
Subjects	Pregnancy--Nutritional aspects.
	Mothers--Nutrition.
	Exercise for pregnant women.
Notes	Includes bibliographical references.

Fast food kids: French fries, lunch lines, and social ties	
LCCN	2018285560
Type of material	Book
Personal name	Best, Amy L., 1970- author.
Main title	Fast food kids: French fries, lunch lines, and social ties / Amy L. Best.
Published/Produced	New York: New York University Press, [2017] ©2017
Description	xiv, 245 pages; 24 cm.
ISBN	9781479842704 (hardback)
	1479842702 (hardback)
	9781479802326 (paperback)
	1479802328 (paperback)
LC classification	TX361.Y6 B47 2017
Summary	The book provides a thorough account of the role that food plays in the lives of today's youth, teasing out the many contradictions of food as a cultural object, fast food portrayed as a necessity for the poor and yet, reviled by upper-middle class parents; fast food restaurants as one of the few spaces that kids can claim and effectively 'take over' for several hours each day; food corporations spending millions each year to market their food to kids and to lobby Congress against regulations; schools struggling to deliver healthy food young people will actually eat, and the difficulty of arranging family dinners, which are known to promote family cohesion and stability. -- amazon.com
Contents	Introduction: Fast food kids -- The family meal: eating together, eating apart -- The cafeteria as great equalizer: making food good -- The cafeteria as youth space: social bonds and

	barriers -- Eat what's good for you: class and the cult of health -- I'm lovin' it: fast food and after-school hot spots -- Conclusion: Food futures and social change -- Methods appendix.
Subjects	Youth--Nutrition--United States.
	Food--Social aspects--United States.
	Convenience foods--Social aspects--United States.
	Youth--United States--Social conditions.
	Food habits--United States.
	Convenience foods--Social aspects.
	Food habits.
	Food--Social aspects.
	Youth--Nutrition.
	Youth--Social conditions.
	United States.
Notes	Includes bibliographical references (pages 187-234) and index.

Food and drink in antiquity: readings from the Graeco-Roman World: a sourcebook

LCCN	2014009059
Type of material	Book
Personal name	Donahue, John F., 1958-
Main title	Food and drink in antiquity: readings from the Graeco-Roman World: a sourcebook / John F. Donahue.
Published/Produced	London; New York: Bloomsbury Academic, 2015.
Description	x, 299 pages: illustrations; 24 cm.
ISBN	9781441133458 (paperback)
	9781441196804 (hardback)
LC classification	GT2853.G8 D66 2015

Summary	"Interest in food and drink as an academic discipline has been growing significantly in recent years. This sourcebook is a unique asset to many courses on food as it offers a thematic approach to eating and drinking in antiquity. For classics courses focusing on ancient social history to introductory courses on the history of food and drink, as well as those offerings with a strong sociological or anthropological approach this volume provides an unparalleled compilation of essential source material. The chronological scope of the excerpts extends from Homer in the Eighth Century BCE to the Roman emperor Constantine in the Fourth Century CE. Each thematic chapter consists of an introduction along with a bibliography of suggested readings. Translated excerpts are then presented accompanied by an explanatory background paragraph identifying the author and context of each passage. Most of the evidence is literary, but additional sources - inscriptional, legal and religious - are also included"-- Provided by publisher.
Contents	Figure List 1. Introduction 2. Food and Drink in Ancient Literature 3. Grain, Grapes and Olives: The Mediterranean Triad and More 4. Eating, Drinking and Believing: Food, Drink and Religion 5. Eating, Drinking and Sharing: The Social Context of Food and Drink 6. Eating, Drinking and Fighting: Food and Drink in the Military 7. Eating, Drinking and Living Healthy: Food, Drink and Medicine Authors and works.
Subjects	Food habits--Greece--History--To 1500.

	Food habits--Rome.
	Civilization, Classical.
	History / Ancient / General.
	History / Social History.
	Social Science / Agriculture & Food.
Notes	"Bloomsbury Sources in Ancient History."
	Includes bibliographical references and index.

Food and health in early modern Europe: diet, medicine and society, 1450-1800	
LCCN	2015004354
Type of material	Book
Personal name	Gentilcore, David.
Main title	Food and health in early modern Europe: diet, medicine and society, 1450-1800 / David Gentilcore.
Published/Produced	London; New York: Bloomsbury Academic, an imprint of Bloomsbury Publishing Plc., 2016.
Description	x, 249 pages: illustrations; 24 cm
ISBN	9781472528896 (HB)
	9781472534972 (PB)
LC classification	RA427.8 .G42 2016
Contents	Healthy food: Renaissance dietetics, c.1450 to c.1650 -- Healthy food: the fall and rise of dietetics, c. 1650 to c. 1800 -- Rich food, poor food: diet, physiology and social rank -- Regional food: nature and nation in Europe -- Holy food: spiritual and bodily health -- Vegetable food: the vegetarian option -- New World food: the Columbian exchange and its European impact -- Liquid food: drinking for health.
Subjects	Health promotion--Europe.

	Nutrition--Europe.
	Food supply--Europe.
	Food consumption--Europe.
Notes	Includes bibliographical references 219-237 and index.

Food and public health: a practical introduction

LCCN	2018006822
Type of material	Book
Uniform title	Food and public health (Karpyn)
Main title	Food and public health: a practical introduction / edited by Allison Karpyn.
Published/Produced	New York: Oxford University Press, [2018]
Description	xv, 368 pages: illustrations; 24 cm
ISBN	9780190626686 (pbk.: alk. paper)
	0190626682 (pbk.: alk. paper)
LC classification	HD9000.6 .F5855 2018
Related names	Karpyn, Allison, editor.
Contents	The history of food and public health / Emily Contois and Anastasia Day -- History and development of the 2015-2020 dietary guidelines for Americans / Alice H. Lichtenstein and Allison Karpyn -- Behavioral design as an emerging theory for dietary behavior change / NCCOR Behavioral Design Working Group -- Health disparities: race, ethnicity, gender and class / Alison G.M. Brown and Sara C. Folta -- Healthy food marketing / Allison Karpyn -- Policy efforts supporting healthy diets for adults and children / Courtney A. Pinard, Eric E. Calloway, Teresa M. Smith, Amy L. Yaroh -- Food insecurity and public health / Molly Knowles, Joanna Simmons,

	Mariana Chilton -- Obesogenic environments and public health mitigation strategies / Allison Karpyn -- Food controversies: the healthy pulse of a democracy? / F. Bailey Norwood -- The obesity pandemic & food insecurity in developing countries: a case study from the Caribbean / Kristen Lowitt, Katherine Gray-Donald, Gordon M. Hickey, Arlette Saint Ville, Isabella Francis-Granderson, Chandra A. Madramootoo, and Leroy Phillip -- Intersections of food and culture: case studies of sugar and meat from Australia, Japan, Thailand, and Nigeria / Wakako Takeda, Cathy Banwell, Kelebogile T. Setiloane, and Melissa K. Melby -- From soil to stomach: agritourism and public health / Erecia Hepburn and Allison Karpyn.
Subjects	Food supply--Government policy--United States.
	Food Industry
	Diet
	Health Policy
	United States
Notes	Includes bibliographical references and index.

Food labels: your questions answered	
LCCN	2019033120
Type of material	Book
Personal name	Brehm-Curtis, Barbara, author.
Main title	Food labels: your questions answered / Barbara A Brehm.
Published/Produced	Santa Barbara, California: ABC-CLIO, 2019.
ISBN	9781440863660 (paperback; acid-free paper)
	(ebook)

LC classification	TP374.5 .B74 2019
Summary	"This book provides an approachable introduction to food labels. While aimed primarily at teens and young adults, it is a valuable tool for anyone who wants to better understand what food labels are really saying and make healthy food choices"-- Provided by publisher.
Subjects	Food--Labeling--Miscellanea.
Notes	Includes bibliographical references and index.
Additional formats	Online version: Brehm-Curtis, Barbara. Food labels Santa Barbara, California: ABC-CLIO, 2019. 9781440863677 (DLC) 2019033121
Series	Q&A health guides

Food labels: your questions answered	
LCCN	2019033121
Type of material	Book
Personal name	Brehm-Curtis, Barbara, author.
Main title	Food labels: your questions answered / Barbara A Brehm.
Published/Produced	Santa Barbara, California: ABC-CLIO, 2019.
Description	1 online resource
ISBN	9781440863677 (ebook)
	(paperback; acid-free paper)
LC classification	TP374.5
Summary	"This book provides an approachable introduction to food labels. While aimed primarily at teens and young adults, it is a valuable tool for anyone who wants to better understand what food labels are really saying and make healthy food choices"-- Provided by publisher.

Subjects	Food--Labeling--Miscellanea.
Notes	Includes bibliographical references and index.
Additional formats	Print version: Brehm-Curtis, Barbara. Food labels Santa Barbara, California: ABC-CLIO, 2019. 9781440863660 (DLC) 2019033120
Series	Q&A health guides

Food oxidants and antioxidants: chemical, biological, and functional properties

LCCN	2012048741
Type of material	Book
Main title	Food oxidants and antioxidants: chemical, biological, and functional properties / edited by Grzegorz Bartosz.
Published/Produced	Boca Raton: CRC Press, [2014]
Description	xvii, 550 pages: illustrations; 24 cm.
ISBN	9781439882412 (hardback)
LC classification	TX553.A3 F64 2014
Related names	Bartosz, G. (Grzegorz)
Summary	"This volume discusses the effects of naturally occurring and process-generated pro-oxidants and antioxidants on various aspects of food quality. It emphasizes the chemical nature and functional properties of these compounds and their interactions with other food components in storage and processing, specifically focusing on the sensory quality, nutritional value, health promoting activity, and safety aspects of foods. It demonstrates the analysis of pro-oxidants and antioxidants in foods, their mechanism and activity, their chemistry and biochemistry, and the practical considerations of healthy food production and marketing"-- Provided by

	publisher.
Subjects	Food additives.
	Oxidizing agents.
	Antioxidants.
	Medical / Nutrition
	Technology & Engineering / Food Science
Notes	Includes bibliographical references and index.
Series	Chemical and functional properties of food components

Food, farms, and community: exploring food systems	
LCCN	2014035951
Type of material	Book
Personal name	Chase, Lisa, author.
Main title	Food, farms, and community: exploring food systems / Lisa Chase & Vern Grubinger.
Published/Produced	Durham, New Hampshire: University of New Hampshire Press, [2014]
Description	288 pages: illustrations; 24 cm
ISBN	9781611684216 (pbk.)
LC classification	HD9000.5 .C488 2014
Related names	Grubinger, Vernon P., 1957- author.
Contents	Introduction to food systems -- Local food systems -- The business of food and farming -- Values in food systems -- The agricultural workforce -- Farming and the environment -- Climate change and agriculture -- Energy, food, and farms -- Access to healthy food -- Farm to school -- Agritourism and on-farm marketing -- Food safety from farm to fork -- The next generation of farmers -- Maintaining farms and farmland for the future -- Improving food systems.

Subjects	Food supply.
	Food security.
Notes	Includes bibliographical references and index.

Good for me. Healthy food

LCCN	2015014972
Type of material	Book
Personal name	Coan, Sharon, author.
Main title	Good for me. Healthy food / Sharon Coan.
Published/Produced	Huntington Beach, CA: Teacher Created Materials, [2016]
Description	12 pages: color illustrations; 20 cm
ISBN	9781493821518 (pbk.)
LC classification	TX355 .C585 2016
Portion of title	Healthy food
Summary	"Your body needs good food. This book will show you some good choices."-- Provided by publisher.
Subjects	Nutrition--Juvenile literature.
	Health--Juvenile literature.
Notes	K to grade 3.
Series	Time For Kids

Hot & hip healthy gluten-free cooking: 75 healthy recipes to spice up your kitchen

LCCN	2015010076
Type of material	Book
Personal name	Matthews, Bonnie, 1963- author.
Main title	Hot & hip healthy gluten-free cooking: 75 healthy recipes to spice up your kitchen / written and photographed by Bonnie Matthews.
Published/Produced	New York, NY: Skyhorse Publishing, [2015]
Description	xi,174 pages; 28 cm

Bibliography

ISBN	9781632202918 (hardback)
LC classification	RM237.86 .M3714 2015
Variant title	Hot and hip healthy gluten-free cooking
Summary	"75 Healthy Recipes to Spice Up Your Kitchen Dozens of recipes that prove gluten-free doesn't mean taste-free. Just because you've gone gluten-free doesn't mean you have to stop eating the foods you love-not even bread, pasta, and dessert! The Badass Gluten-Free Cookbook makes it easy to enjoy all the benefits of a gluten-free diet while indulging in home-cooked meals that are as delicious as they are easy to make. The Badass Gluten-Free Cookbook features a wide variety of wholesome and tempting recipes that will satisfy both your passion for good food and your gluten-free lifestyle: Mesquite flour savory breakfast muffins Grilled panini with buffalo mozzarella, roasted red peppers, and sundried tomatoes Homemade ravioli with yam filling and sage butter Ahi tuna kebabs over buckwheat noodles with miso sauce Crusted chicken with fire-roasted tomato sauce Pear and cranberry crisp Chocolate biscotti with chipotle spice And many more! This book also includes a helpful guide to stocking your kitchen with gluten-free staples and substitutes, and sumptuous full-color photographs that will inspire your inner chef. The Badass Gluten-Free Cookbook is a no-nonsense guide to cooking great, healthy food for your badass, gluten-free life"-- Provided by publisher.

	"Just because you've gone gluten-free doesn't mean you have to stop eating the foods you love--not even bread, pasta, and dessert! The Badass Gluten-Free Cookbook makes it easy to enjoy all the benefits of a gluten-free diet while indulging in home-cooked meals that are as delicious as they are easy to make. The Badass Gluten-Free Cookbook features a wide variety of wholesome and tempting recipes that will satisfy both your passion for good food and your gluten-free lifestyle: Mesquite flour savory breakfast muffins Grilled panini with buffalo mozzarella, roasted red peppers, and sundried tomatoes Homemade ravioli with yam filling and sage butter Ahi tuna kebabs over buckwheat noodles with miso sauce Crusted chicken with fire-roasted tomato sauce Pear and cranberry crisp Chocolate biscotti with chipotle spice And many more! This book also includes a helpful guide to stocking your kitchen with gluten-free staples and substitutes, and sumptuous full-color photographs that will inspire your inner chef. The Badass Gluten-Free Cookbook is a no-nonsense guide to cooking great, healthy food for your badass, gluten-free life"--Provided by publisher.
Subjects	Gluten-free diet--Recipes.

How to choose foods your body will use	
LCCN	2015041590
Type of material	Book
Personal name	Sjonger, Rebecca, author.
Main title	How to choose foods your body will use / Rebecca Sjonger.

Published/Produced	New York, New York: Crabtree Publishing Company, [2016]
Description	24 pages: color illustrations; 25 cm.
ISBN	9780778723509 (reinforced library binding)
	9780778723523 (pbk.)
LC classification	TX355 .S5967 2016
Summary	"With a focus on building health and nutrition literacy, this timely title gives readers the tools they need to make healthy food choices for every meal! Topics include whole foods, processed foods, and reading food labels. Healthful food tips are explained using text that is easy for young readers to digest"-- Provided by publisher.
Contents	Healthy habits -- Whole foods -- Processed foods -- Junk foods -- Energy in, energy out -- Meals and snacks -- Mix it up! -- Tasty tips -- What's for dinner? -- Show what you know! -- Learning more -- Words to know and index.
Subjects	Nutrition--Juvenile literature.
	Diet--Juvenile literature.
	Health--Juvenile literature.
Notes	Includes index.
	Ages 5-8.
	K to grade 3.
Series	Healthy habits for a lifetime

Innovations in sustainability: fuel and food	
LCCN	2015490085
Type of material	Book
Personal name	Marcus, Alfred Allen, 1950- author.
Main title	Innovations in sustainability: fuel and food / Alfred A. Marcus, University of Minnesota.

Published/Produced	Cambridge, U.K.: Cambridge University Press, 2015.
Description	xvii, 363 pages; 23 cm
Links	Contributor biographical information http://www.loc.gov/catdir/enhancements/fy1606/2015490085-b.html
	Publisher description http://www.loc.gov/catdir/enhancements/fy1606/2015490085-d.html
	Table of contents http://www.loc.gov/catdir/enhancements/fy1606/2015490085-t.html
ISBN	9781107072794 (hb.)
	1107072794 (hb.)
	9781107421110 (pbk.)
	110742111X (pbk.)
LC classification	HD30.255 .M344 2015
Summary	"To what extent can competition between companies encourage innovations in sustainability that have the potential to solve some of the world's major challenges? Using a series of case studies, this book pits closely related competitors against each other to examine the progress in and obstacles to the evolution of sustainable innovations in energy efficiency, solar power, electric vehicles and hybrids, wind energy, healthy eating, and agricultural productivity. It delves into the efforts of Tesla Motors to bring about a revolution in personal transportation, and the challenges Toyota and General Motors (GM) confront in commercializing hybrids. It explores the movement to healthy food by cereal companies General Mills and Kellogg's, and depicts the battles between Whole Foods and

Bibliography 111

	Walmart for the world's palate..."--Page 4 of cover.
Contents	Introduction: the path to sustainability. I. Funding sustainable startups. Leaders of the pack: Khosla Ventures and KPCB -- Scaling up: Intel Capital and Google Ventures -- II. Business models. Follow the sun: First Solar and Suntech -- Making a revolution: Tesla and Better Place -- III. The macroenvironment and industry context. Ticket to ride: Toyota and General Motors -- Blowing in the wind: Vestas and General Electric -- IV. Finding customers. Carrying that weight: General Mills and Kellogg's -- Bridge over troubled waters: Pepsi and Coca-Cola -- V. Competition between mission and non-mission based business. Consensus capitalism: Whole Foods and Walmart -- Sustainability's next frontier: DuPont and Monsanto -- Concluding observations: the journey continues.
Subjects	Industrial management--Environmental aspects.
	Technological innovations.
	Green products.
	Sustainable development.
	Strategic planning.
	Competition.
Notes	Includes bibliographical references and index.
Series	Organizations and the natural environment

It's not about the broccoli: three habits to teach your kids for a lifetime of healthy eating

LCCN	2013033502
Type of material	Book

Personal name	Rose, Dina, author.
Main title	It's not about the broccoli: three habits to teach your kids for a lifetime of healthy eating / Dina Rose, PhD.
Edition	First edition.
Published/Produced	New York, New York: Perigee, 2014.
Description	viii, 256 pages: illustrations; 21 cm
ISBN	9780399164187 (pbk.)
LC classification	RJ206 .R668 2014
Summary	"You already know how to give your kids healthy food. But the hard part is getting them to eat it. After years of research and working with parents, Dina Rose discovered a powerful truth: When parents focus solely on nutrition, their kids--surprisingly--eat poorly. But when families shift their emphasis to behaviors--the skills and habits kids are taught--they learn to eat right. Every child can learn to eat well--but only if you show them how to do it. Dr. Rose describes the three habits--proportion, variety, and moderation--all kids need to learn, and gives you clever, practical ways to teach these food skills. All children can learn: - How to confidently explore strange, new foods - How to know when they're hungry and when they're full - What to do when they say they're 'starving'--and about to attend a birthday party - How to branch out from easy-to-like prepackaged kid fare to more mature tastes and textures: savory, tangy, runny, crunchy - How to engage in open and honest talk about food without yelling, "I don't like it!" With It's Not About the Broccoli, you can teach your children how to eat, and give them the skills they need for a lifetime of

	health and vitality"-- Provided by publisher.
Subjects	Children--Nutrition.
	Natural foods.
	Health promotion.
	Family & Relationships / Parenting / General.
Notes	Includes bibliographical references (pages 245-254).

Kraft	
LCCN	2015049958
Type of material	Book
Personal name	Green, Sara, 1964- author.
Main title	Kraft / by Sara Green.
Published/Produced	Minneapolis, MN: Bellwether Media, Inc., 2017.
Description	24 pages: illustrations; 24 cm.
ISBN	9781626174092 (hardcover: alk. paper)
LC classification	HD9009.K73 G74 2017
Summary	"Engaging images accompany information about Kraft. The combination of high-interest subject matter and narrative text is intended for students in grades 3 through 7"-- Provided by publisher.
Contents	What is Kraft? -- How Kraft began -- Growth and change -- Fun and healthy food -- Kraft gives back -- Kraft timeline -- Glossary -- To learn more -- Index.
Subjects	Kraft, James Lewis, 1874-1953--Juvenile literature.
	Kraft Heinz Company--History--Juvenile literature.
	Food industry and trade--United States--History--Juvenile literature.

Notes	Includes bibliographical references and index.
	Age 7-13.
Series	Pilot. Brands we know

Living with a green heart: how to keep your body, your home, and the planet healthy in a toxic world

LCCN	2018289944
Type of material	Book
Personal name	Browne, Gay, author.
Main title	Living with a green heart: how to keep your body, your home, and the planet healthy in a toxic world / Gay Browne.
Published/Produced	New York, NY: Citadel Press/Kensington Publishing Corp., [2019]
	©2019
Description	xiii, 274 pages; 21 cm
ISBN	9780806539003 (paperback)
	0806539003 (paperback)
LC classification	GE196 .B76 2019
Summary	"Is the damage we're doing to our planet literally leaving you sick, sore, and gasping for air? What to take back our inalienable rights to clean air, clean water, and healthy food? In this quietly revolutionary book, environmental pioneer and founder of Greenopia, Gay Browne, shares a roadmap for making incremental changes that will not only transform your life, but heal the world we share"--Page [4] of cover.
Contents	Foreword / by Terry Tamminen -- Introduction: My green story -- The elements: air, water, and EMFs. The air that you breathe; Water is life; EMFs -- Personal care. The truth about toothpaste; Deodorant; Soap; Hair care; Sun

Bibliography

	protection; Makeup and moisturizers; Intimate care products -- Food and beverages. Dairy; Red meat; Poultry; Eggs; Something funny about fish; Fruits and vegetables; Grains, flours, and pasta; Plant-based proteins; Nuts and seeds; Sugar and spice and everything nice (and not so nice); Coffee and tea; Alcoholic beverages; Nonalcoholic beverages -- Home. Household cleaning products; The family room; The bedroom; The kitchen; The bathroom; The garage; Waste disposal -- Lifestyle. Pets; Clothing; The real cost of transportation; Socially responsible investing and spending -- Get involved -- So, what do I do now?
Subjects	Sustainable living.
	Toxins.
Notes	Includes index.

Lunch	
LCCN	2013015689
Type of material	Book
Personal name	Parker, Victoria, author.
Main title	Lunch / Vic Parker.
Published/Produced	Chicago, Illinois: Heinemann Library, [2014]
Description	32 pages: color illustrations; 23 cm.
ISBN	9781432991173 (hb)
	1432991175 (hb)
	9781432991227 (pb)
	1432991221 (pb)
LC classification	TX355 .P2547 2014
Summary	"Read Lunch to learn how to make healthy food choices during this midday meal. Different photos show healthy and unhealthy lunch

	options, while simple text explains why some choices are better than others. A lunch foods quiz concludes the book."-- Provided by publisher.
Contents	Why make healthy choices? -- What makes a lunch healthy or unhealthy? -- Savory sandwiches -- Sweet sandwiches -- Burgers -- Hot dogs -- Chicken -- Noodles and pasta -- Soup -- Salads -- Drinks -- Food quiz -- Food quiz answers -- Tips for healthy eating.
Subjects	Nutrition--Juvenile literature.
	Lunchbox cooking--Juvenile literature.
	Health--Juvenile literature.
Notes	Includes bibliographical references (page 32) and index.
Series	Healthy food choices
	Heinemann first library

Mad delicious: the science of making healthy food taste amazing!	
LCCN	2014938999
Type of material	Book
Personal name	Schroeder, Keith, author.
Main title	Mad delicious: the science of making healthy food taste amazing! / Keith Schroeder.
Published/Produced	New York, New York: Oxmoor House, [2014] ©2014
Description	384 pages: color illustrations; 24 cm
Links	Contributor biographical information https://www.loc.gov/catdir/enhancements/fy1702/2014938999-b.html
	Publisher description https://www.loc.gov/catdir/enhancements/fy1702/2014938999-d.html
ISBN	9780848704285

	0848704282
LC classification	TX651 .S373 2014
Variant title	At head of title: Cooking light
Related titles	Cooking light (Oxmoor House)
Summary	Mad Delicious takes the kitchen science genre to the next level: It's not just about chemistry and molecules. Schroeder teaches home cooks about the nature of ingredients, how to maximize texture and flavor with clever cooking techniques (try steaming beef-then soaking it in wine sauce for the most tender steak ever!), smooth moves in the kitchen for better work flow, and how all the sciences-geography, meteorology, chemistry, physics, botany, biology, even human sociology and anthropology-can help home cooks master the science of light cooking.
Contents	Introduction -- Mental mise en place: how to get your kitchen and mind ready for cooking -- Hands on: master basic kitchen skills - Sauces & dressings: how to create sauces with fresh flavors and great texture -- Hot liquids: use steam for sides and main dishes that wow -- Harnessing steam: all about poaching, simmering, and boiling -- Pan cooking: sautéing, frying, braising, and more -- In the over: how to roast anything: nuts, veggies, meat, and fruit -- Playing with fire: recipes for the grill and the grill pan.
Subjects	Cooking.
Notes	"126 crazy-good recipes!"--Cover.
	Includes indexes.

Make your own rules diet	
LCCN	2014012500
Type of material	Book
Personal name	Stiles, Tara.
Main title	Make your own rules diet / Tara Stiles.
Published/Produced	Carlsbad: Hay House, Inc., [2014]
Description	xxi, 285 pages: color illustrations; 24 cm
ISBN	9781401944353 (hardback)
LC classification	BF632 .S75 2014
Summary	"In Make Your Own Rules Diet, Tara Stiles introduces readers to easy and fun ways to bring yoga, meditation, and healthy food into their lives. As the designer and face of Reebok's first yoga lifestyle line, author of Yoga Cures and Slim, Calm, Sexy Yoga, and the founder of Strala--the movement-based system that ignites freedom, known for its laid-back and unpretentious vibe--Tara has long been a proponent of creating a tension-free healthy life by tapping into the unique needs of her clients. In this new book, she teaches readers how to apply this inward-looking philosophy to themselves. When people understand what they need for true well-being, they can make their own rules--rules that will help them become their best selves. In her rulebook, it's no pain, much gain. In fact, Tara stresses the importance of practicing with ease--leaving the discomfort and tension behind--because what you practice is what you manifest. Readers will not only learn to create their own rules but also to understand when something isn't working anymore, so they can update their rules as

Bibliography

	circumstances change. Her approach takes readers from the kitchen, to the mat, to the cushion, in an effort to help them get to know themselves. After leading them through some basic guidelines about how to write their rulebooks, Tara lays out tips, techniques, and practices, including; A step-by-step goal setting process so readers can figure out where they want to focus; Six yoga routines specifically designed to up energy levels, curb cravings, drop pounds, and enhance peace; Eight breathing and meditation practices to soothe the soul; 50 simple, delicious, plant-based recipes that can be made in minutes; A 7-day kick-start program and a 30-day transformation plan to launch readers on their healthy, happy, radiant path. So join Tara today as she opens readers' eyes to a new way of living well that anyone can do--no matter where they are now. "-- Provided by publisher.
Subjects	Self-control.
	Yoga.
	Diet.
	Health & Fitness / Diets.

Measurement of antioxidant activity and capacity: recent trends and applications	
LCCN	2017059633
Type of material	Book
Main title	Measurement of antioxidant activity and capacity: recent trends and applications / edited by Reşat Apak, Istanbul University, Esra Çapanoğlu, Istanbul Technical University,

	Fereidoon Shahidi, Memorial University of Newfoundland.
Edition	First edition.
Published/Produced	Hoboken, NJ, USA: Wiley, 2018.
Description	1 online resource.
ISBN	9781119135364 (pdf)
	9781119135371 (epub)
LC classification	TX553.A73
Related names	Apak, Reşat, editor.
	Çapanoğlu, Esra, editor.
	Shahidi, Fereidoon, 1951- editor.
Summary	"A comprehensive reference for assessing the antioxidant potential of foods and essential techniques for developing healthy food products Measurement of Antioxidant Activity and Capacity offers a much-needed resource for assessing the antioxidant potential of food and includes proven approaches for creating healthy food products. With contributions from world-class experts in the field, the text presents the general mechanisms underlying the various assessments, the types of molecules detected, and the key advantages and disadvantages of each method. Both thermodynamic (i.e., efficiency of scavenging reactive species) and kinetic (i.e., rates of hydrogen atom or electron transfer reactions) aspects of available methods are discussed in detail. A thorough description of all available methods provides a basis and rationale for developing standardized antioxidant capacity/activity methods for food and nutraceutical sciences and industries. This text also contains data on new antioxidant

	measurement techniques including nanotechnological methods in spectroscopy and electrochemistry, as well as on innovative assays combining several principles. Therefore, the comparison of conventional methods versus novel approaches is made possible. This important resource: Offers suggestions for assessing the antioxidant potential of foods and their components Includes strategies for the development of healthy functional food products Contains information for identifying antioxidant activity in the body Presents the pros and cons of the available antioxidant determination methods, and helps in the selection of the most appropriate method. Written for researchers and professionals in the nutraceutical and functional food industries, academia and government laboratories, this text includes the most current knowledge in order to form a common language between research groups and to contribute to the solution of critical problems existing for all researchers working in this field."-- Provided by publisher.
Subjects	Food--Analysis.
	Antioxidants.
	Technology & Engineering / Food Science.
Notes	Includes bibliographical references and index.
Additional formats	Print version: Measurement of antioxidant activity and capacity First edition. Hoboken, NJ, USA: Wiley, 2018 9781119135357 (DLC) 2017058596

Natural resources and technology	
LCCN	2016323412

Type of material	Book
Main title	Natural resources and technology / editor, Darlina Md. Naim, Mardiana Idayu Ahmad.
Published/Produced	[Kampung Sungai Glugur], Pulau Pinang: Penerbit Universiti Sains Malaysia, [2015]
	[Kampung Sungai Glugur], Pulau Pinang, Malaysia: Universiti Sains Malaysia Co-operative Bookshop Ltd.
Description	x, 66 pages; 25 cm
ISBN	9789838619172
LC classification	HC445.5.Z65 N38 2015
Related names	Darlina Md. Naim, 1979- editor.
	Mardiana Idayu Ahmad, 1980- editor.
	Penerbit Universiti Sains Malaysia, publisher.
Contents	Rimbo larangan, a natural conservation and sustainable forest management practices of Minangkabau / Onrizal, Mashhor Mansor -- Preparation of local bamboo charcoal via traditional method and its potential as primary humidity absorber / Ezzah Azimah Alias, Norzaini Zainal, Mardiana Idayu Ahmad -- Rice cultivation system in Malaysia, from traditional to contemporary and its weed problems in paddy filed / Siti Norasikin Ismail, Mashhor Mansor -- Kelantanese dishes, authenticity versus traditional healthy food / Mohd. Yusof Zulkefli, Nor Hazlina Hashim -- Plants as bio-indicators based on air pollution tolerance index (APTI) approach / Siti Noor Aisyah Mohd. Sabri, Mardiana Idayu Ahmad, Darlina Md. Naim -- Malay house ventilation, past, and present / Fatin Zafirah Mansur, Siti Masitah Abul Rahman, Mardiana Idayu Ahmad --

	Turbine frame and float design for micro hydro turbine at Lepung Gaat, Kapit, Sarawak / Saiful Bahari Mohd. Yusoff.
Subjects	Natural resources--Malaysia.
	Natural resources--Technological innovations--Malaysia.
Notes	Includes bibliographical references and index.
	In English.

Orthorexia: when healthy eating goes bad

LCCN	2017289443
Type of material	Book
Personal name	McGregor, Renee, author.
Main title	Orthorexia: when healthy eating goes bad / Renee McGregor, RD.
Published/Produced	London: Nourish Books, 2017.
Description	xii, 210 pages: illustrations; 20 cm
ISBN	184899334X (paperback)
	9781848993341 (paperback)
LC classification	RC552.E18 M412 2017
Related names	Wilson, Bee, writer of foreword.
Summary	Orthorexia is an unhealthy obsession with eating only healthy food. It is closely related to anorexia, but focused on quality of food rather than quantity. But how do you know if you or a friend or loved one has crossed that line? And how much do you really know about the impact these diets, plans and detoxes are having on your body?-- Source other than the Library of Congress.
Subjects	Eating disorders.
Notes	Includes index.

Pumpkin soup and cherry bread: a Steiner-Waldorf Kindergarten cookbook	
LCCN	2018487199
Type of material	Book
Personal name	Rosengren, Rikke, author.
Uniform title	Mad og nærvær. English
Main title	Pumpkin soup and cherry bread: a Steiner-Waldorf Kindergarten cookbook / Rikke Rosengren and Nana Lyzet; photographs by Stine Heilmann; translated by Agnes Broome.
Published/Produced	Edinburgh: Floris Books, 2015.
Description	167 pages: color illustrations; 23 cm
ISBN	9781782502005 (pbk.)
	1782502009 (pbk.)
LC classification	TX714 .R6733 2015
Related names	Lyzet, Nana, author.
	Heilmann, Stine, photographer.
Summary	Contains over eighty seasonal recipes -- nutritious, delicious, vegetarian and 100% natural -- along with a wealth of advice on encouraging children to enjoy healthy food.
Subjects	Cooking.
	Cooking (Vegetables)
	Children--Nutrition.
	Cooking--Study and teaching (Preschool)--Activity programs.
	Waldorf method of education.
Notes	Originally published in Danish as: Mad og nærvær, 2014.
	Includes index.

Quitting plastic: easy and practical ways to cut down the plastic in your life	
LCCN	2018493995
Type of material	Book
Personal name	Roldan, Clara Williams, author.
Main title	Quitting plastic: easy and practical ways to cut down the plastic in your life / Clara Williams Roldan; with Louise Williams; illustrations by Elowyn Williams Roldan.
Published/Produced	Sydney, NSW: Allen & Unwin, 2019.
	©2018
Description	216 pages: illustrations; 18 cm
ISBN	9781760528713 (paperback)
LC classification	TD798 .R65 2019
Related names	Williams, Louise, 1961- author.
	Roldan, Elowyn Williams, illustrator.
Summary	Where do you start if you want to reduce the plastic in your life? Especially when most of us are wearing it, eating and drinking from it, sitting on it, walking on it, and probably even ingesting it. Anywhere you go, plastic is within easy reach - even in Antarctica and the North Pole. We didn't quit plastic overnight. In fact, it's still a work in progress. But along the way, we have learnt a lot - researching the issue from the grass roots up, speaking to people and finding out what works and what doesn't. We take on the tricky questions, like 'how will I wash my hair?', 'do I have to give up crackers?', 'what about my bin liner?' and 'is this going to be expensive?' As we continue to remove throw-away plastics from our daily lives, we've discovered we're friendlier with our local

	communities, we're eating more healthy food, and de-cluttering happens by itself. It feels great!
Contents	Breaking up with plastic -- Good plastic, bad plastic -- Old habits die hard -- Getting started -- The kitchen -- The laundry and cleaning -- The bathroom -- Your wardrobe -- Plastic-free kids -- Entertaining, and eating (and drinking) out -- The end of single-use plastic? -- What more can I do? -- Acknowledgements -- Notes.
Subjects	Plastic scrap--Recycling.
	Plastic scrap--Environmental aspects.
	Environmental protection--Citizen participation--Handbooks, manuals, etc.
	Waste minimization--Handbooks, manuals, etc.
	Sustainable living--Handbooks, manuals, etc.
	Plastics industry and trade--Environmental aspects.
Notes	Includes bibliographical references.

Real southern barbecue: constructing authenticity in southern food culture

LCCN	2019021530
Type of material	Book
Personal name	Byrd, Kaitland M., author.
Main title	Real southern barbecue: constructing authenticity in southern food culture / Kaitland M. Byrd.
Published/Produced	Lanham: Lexington Books, [2019]
Description	1 online resource.
ISBN	9781498593366 (Electronic)
LC classification	TX840.B3
Contents	Eating authentic food: producing and

	consuming authenticity in food culture -- The meaning of barbecue: the role of place, style, and tradition -- Beyond the meat and sauce: barbecue as a healthy food option -- Real barbecue restaurant have smoke: the impact of environmental concerns on barbecue techniques -- Family, commercialization, and community dynamics: the future of barbecue -- The new taste: failed authenticity and changing taste in food culture.
Subjects	Barbecuing.
	Barbecuing--Southern States.
	Cooking, American--Southern style.
Form/Genre	Cookbooks.
Notes	Includes bibliographical references and index.
Additional formats	Print version: Byrd, Kaitland M., author. Real southern barbecue Lanham: Lexington Books, [2019] 9781498593359 (DLC) 2019020302

Real southern barbecue: constructing authenticity in southern food culture

LCCN	2019020302
Type of material	Book
Personal name	Byrd, Kaitland M., author.
Main title	Real southern barbecue: constructing authenticity in southern food culture / Kaitland M. Byrd.
Published/Produced	Lanham: Lexington Books, [2019]
ISBN	9781498593359 (cloth: alk. paper)
LC classification	TX840.B3 B93 2019
Contents	Eating authentic food: producing and consuming authenticity in food culture -- The meaning of barbecue: the role of place, style,

	and tradition -- Beyond the meat and sauce: barbecue as a healthy food option -- Real barbecue restaurant have smoke: the impact of environmental concerns on barbecue techniques -- Family, commercialization, and community dynamics: the future of barbecue -- The new taste: failed authenticity and changing taste in food culture.
Subjects	Barbecuing.
	Barbecuing--Southern States.
	Cooking, American--Southern style.
Form/Genre	Cookbooks.
Notes	Includes bibliographical references and index.
Additional formats	Online version: Byrd, Kaitland M., author. Real southern barbecue Lanham: Lexington Books, [2019] 9781498593366 (DLC) 2019021530

Research handbook of marketing in emerging economies	
LCCN	2016959913
Type of material	Book
Main title	Research handbook of marketing in emerging economies / edited by Marin A. Marinov, Professor of International Business, Aalborg University, Denmark.
Published/Produced	Cheltenham, UK; Northampton, MA, USA: Edward Elgar Publishing, [2017]
	©2017
Description	xvi, 310 pages: illustrations; 24 cm
ISBN	1784713163
	9781784713164
LC classification	HF5415.12.D44 R47 2017
Related names	Marinov, Marin, 1948- editor.
Contents	Introduction: Marketing in emerging economies

| | / Marin A. Marinov -- 1 Data collection procedure equivalence in emerging economy market research / Pervez N. Ghauri, Agnieszka Childlow -- 2 Globalization, sustainability and marketing of healthcare in emerging markets: Doing good while doing well / Van R. Wood -- 3 Marketing accountability in emerging economy firms / Maja Arslangić-Kalajdžić, Vesna Žabkar -- 4 Materialistic tendencies and adolescent healthy food consumption: Setting the research agenda / Nesma Ammar, Noha El-Bassiouny, Ronia Hawash -- 5 Psychobranding of emerging economy firms: Building emotional connections with local consumers / G. Nicolás Kfuri -- 6 Multinational corporation retailing in emerging economies: Interplays of resistance, cooperation and transmutation / Marie-Laure Baron, Ruby Roy Dholakia, Nikhilesh Dholakia -- 7 Perceived advertising intrusiveness and avoidance in emerging economies: The case of China / Dan A. Petrovici, Svetla T. Marinova, Marin A. Marinov -- 8 Value branding in emerging economies as a social dimension in the Indian context / S. Ramesh Kumar, Svetla T. Marinova -- 9 Researching country image construct in the context of emerging economies / Durdana Ozretic-Dosen, Vatroslave Skare, Zoran Krupka -- 10 Opening the black box of Russian culture in B2B relationships / Carl Arthur Solberg, Anzhelika Osmanova -- 11 Russian consumer behavior: In search of a balance between national uniqueness and Western mainstream / Sergei F. Sutyrin, Irina V. Vorobieva -- 12 Marketing in an emerging |

	economy: The Russian e-commerce market / Maria Smirnova, Vera Rebiazina, Anna Daviy -- 13 Marketing in Bulgaria: A small emerging economy and multicultural markets / Vesselin Blagoev, Michael Minkov -- 14 Diffusion of supermarkets in Bangladesh: Miles to go / M. Yunus Ali, Anisur Rahman Faroque.
Subjects	Marketing--Developing countries.
Notes	Includes bibliographical references and index.
Additional formats	Online version: Research handbook of marketing in emerging economies. Cheltenham, Glos: Edward Elgar Publishing Limited; Northampton, Massachusetts: Edward Elgar Publishing, Inc., [2017] 9781784713171 (OCoLC)983345224
Series	Research handbooks in business and management

S.N.A.C. it up!: Shi's natural approaches to cooking.

LCCN	2017300181
Type of material	Book
Personal name	Curry, Shiona Shi, author.
Main title	S.N.A.C. it up!: Shi's natural approaches to cooking.
Published/Produced	[United States]: Spirit Reign Publishing, [2016] 2016
Description	120 pages: color illustrations; 28 cm
ISBN	9781940002798 (softback)
LC classification	TX833.5 .C87 2016
Cover title	Cookbook with "Shi"
Summary	"Shiona "Shi" Curry is the 11 year old CEO of S.N.A.C. it up! (Shi's Natural Approach to Cooking). Her passion for eating healthy has

	translated into a desire for all children to be healthy while enjoying their food"-- Provided by publisher.
Contents	Testimonials -- Welcome -- "Shi's kitchen commandments" -- What do we need in order to be healthy? -- Common kitchen Measurements -- The healthy food pyramid -- Granny's cinnamon apple oatmeal -- Breakfast hash browns -- Yummy raw granola bars -- Breakfast sausage muffins -- Almond or peanut -- Butter oatmeal muffins -- Simple summer whipped fruit pie -- No bake healthy brownies -- Honeydew lemon cupcakes -- Marshmallow ice cream cake -- Frozen strawberry raspberry yogurt pops -- Sassy applesauce -- Cucarrot salad -- Carob joy-nut bars -- Apple peanut butter cup -- Shi-trail mix -- Apple zinger juice -- Ginger lemonade -- Healthy sugar juice -- Gotta have my greens juice? -- Puzzles -- Thank you's.
Subjects	Quick and easy cooking.
	Children--Nutrition.

Science, democracy, and a healthy food policy	
LCCN	2015304809
Type of material	Book
Personal name	Bailin, Deborah, author.
Main title	Science, democracy, and a healthy food policy / Deborah Bailin, Pallavi Phartiyal.
Published/Produced	Cambridge, MA: Union of Concerned Scientists, 2014.
Description	27 pages: color illustrations; 28 cm
LC classification	MLCM 2018/49587 (T)

Related names	Phartiyal, Pallavi, author.
	Union of Concerned Scientists.
Subjects	Nutrition policy
Notes	Cover title.
	"A Lewis M. Branscomb Science and Democrary Forum."

Seven grains of paradise: a culinary journey in Africa

LCCN	2017430605
Type of material	Book
Personal name	Baxter, Joan, author.
Main title	Seven grains of paradise: a culinary journey in Africa / Joan Baxter.
Published/Produced	Lawrencetown Beach, Nova Scotia, Canada: Pottersfield Press, [2017]
	©2017
Description	285 pages: illustrations; 23 cm
ISBN	9781988286020 (paperback)
LC classification	GT2853.A35 B39 2017
Variant title	7 grains of paradise
Summary	"Seven grains of paradise tells the fascinating and much neglected story about many kinds of food--and also delicacies--in Africa, a continent that gets precious little credit for anything, least of all its intricate cuisines, farms, farming know-how, food cultures and its ability to feed itself. It shouldn't be surprising that Africa has all of these, but for many it may be. Centuries of disparaging judgements and a half century of media reports churning out images of famine, disease and conflict on the continent, have eclipsed the facts that Africans have marvellous local foods and culinary cultures, and that small

family farms still feed most of the continent. The narrative of the book is driven by Baxter's personal quest to learn about some fascinating and new (to her) foods in a handful of countries in sub-Sahara Africa and collect the stories these tell about the continent's farms, its markets, and its people. Her guides and tutors are the people who grow, sell, buy, prepare, and serve the foods. They help her explore the riddles of a continent better known for hunger than for its food, and why that is. It draws on stories collected over the more than thirty years that she has lived and worked in Africa, and builds on these with meticulous research. From the fabled city of Timbuktu on the southern edge of the Sahara Desert, to the diamond fields of Sierra Leone, from the savannah of northern Ghana, to the rainforests of Central Africa, readers are invited along on a delightful journey of learning and eating--and some drinking too, of invigorating indigenous beverages, brews and palm wine straight from the trees. The culinary journey takes the reader down garden paths, into forests that double as farms, through the chaos of markets and into modest little roadside eateries. The real surprise here is not that Africa should boast such a dazzling array of delicious dishes and culinary traditions and indigenous foods; it is that the rest of the world knows so little about them. Baxter, a journalist, anthropologist, development researcher and writer, and Senior Fellow with the independent think tank, the Oakland Institute, does not shy away from the realities of hunger and poverty

	and the real lack of amenities, health facilities, and sanitation on the continent. While the book highlights the complexities and delights of African foods and family farms, it also documents the growing risks they face. So even if Seven grains of paradise is intimate and often light in tone, it is also an important and eye-opening work, thoroughly researched. The stories feed the overarching narrative of what makes for healthy food and farms and communities--what they are and how to maintain them on a continent where "slow food" and "local food" are still the normal fare for so many. With its focus on food, the book is timely and likely to garner much attention as the world confronts rising food prices, and the future of food--and the farms that grow it--in the face of climate change."-- Provided by publisher.
Subjects	Baxter, Joan--Travel--Africa, Sub-Saharan.
	Food--Africa, Sub-Saharan.
	Local foods--Africa, Sub-Saharan.
	Food habits--Africa, Sub-Saharan.
	Cooking--Africa, Sub-Saharan.
	Farms, Small--Africa, Sub-Saharan.
	Africa, Sub-Saharan--Description and travel.
Notes	Includes bibliographical references.

Snacks	
LCCN	2013015692
Type of material	Book
Personal name	Parker, Victoria, author.
Main title	Snacks / Vic Parker.
Published/Produced	Chicago, Illinois: Heinemann Library, [2014]

Description	32 pages: color illustrations; 23 cm.
ISBN	9781432991197 (hb)
	1432991191 (hb)
	9781432991241 (pb)
	1432991248 (pb)
LC classification	TX355 .P255 2014
Summary	"Read Snacks to learn how to make healthy food choices at snack time. Different photos show healthy and unhealthy snack options, while simple text explains why some choices are better than others. A snack foods quiz concludes the book."-- Provided by publisher.
Contents	Why make healthy choices? -- What makes a snack healthy or unhealthy? -- Fruit -- Cheese -- Cookies -- Popcorn -- Chips -- Yogurt -- Dips -- Sweet treats -- Drinks -- Food quiz -- Food quiz answers -- Tips for healthy eating.
Subjects	Nutrition--Juvenile literature.
	Snack foods--Juvenile literature.
	Health--Juvenile literature.
Notes	Includes bibliographical references (page 32) and index.
Series	Healthy food choices
	Heinemann first library

Sustainable food and agriculture: an integrated approach	
LCCN	2019301192
Type of material	Book
Main title	Sustainable food and agriculture: an integrated approach / edited by Clayton Campanhola, Shivaji Pandey.
Published/Produced	London: Academic Press, is an imprint of Elsevier, 2019.

Description	xxiii, 585 pages: illustrations, maps; 24 cm
ISBN	9780128121344 (pbk.)
	0128121343 (pbk.)
LC classification	HD9000.5 .S832 2019
Related names	Campanhola, Clayton, editor.
	Pandey, Shivaji, editor.
Summary	Sustainable Food and Agriculture: An Integrated Approach is the first book to look at the imminent threats to sustainable food security through a cross-sectoral lens. As the world faces food supply challenges posed by the declining growth rate of agricultural productivity, accelerated deterioration of quantity and quality of natural resources that underpin agricultural production, climate change, and hunger, poverty and malnutrition, a multi-faced understanding is key to identifying practical solutions. This book gives stakeholders a common vision, concept and methods that are based on proven and widely agreed strategies for continuous improvement in sustainability at different scales. While information on policies and technologies that would enhance productivity and sustainability of individual agricultural sectors is available to some extent, literature is practically devoid of information and experiences for countries and communities considering a comprehensive approach (cross-sectoral policies, strategies and technologies) to SFA. This book is the first effort to fill this gap, providing information on proven options for enhancing productivity, profitability, equity and environmental sustainability of individual sectors and, in

	addition, how to identify opportunities and actions for exploiting cross-sectoral synergies.
Contents	1 Food and Agricultural Systems at a Crossroads: An Overview; 1.1 The State of Global Food and Agriculture; 1.2 Food and Agriculture at a Crossroads: Challenges and Opportunities; Global Trends and Challenges to Food and Agriculture Into the 21st Century; The Demographics of Rural Poverty and Sustainable Agriculture and Rural Transformations Climate Change, Agriculture, and Food Security: Impacts and the Potential for Adaptation and Mitigation Water Scarcity and Challenges to Food Security; Forests, Land-Use Change, and Challenges to Climate Stability and Food Security; Land and Water Governance, Poverty, and Sustainability; Biodiversity and Ecosystem Services; Changing Food Systems: Implications for Food Security and Nutrition; 1.3 Global Challenges, Global Responses; 2 Global Trends and Challenges to Food and Agriculture into the 21st Century; 2.1 Introduction; 2.2 Key Trends and Challenges Agricultural Demand is Expected to Increase Significantly to Meet the Demand, Output Will Have to Expand, but Under Increasingly Tight Production Constraints; Climate Change Adds Another Major Challenge; Successes in Hunger and Poverty Reduction are Undeniable, But Much More Needs to be Done; Food Systems are Changing, Jeopardizing many Landless and Smallholder Farmers; 2.3 Key Questions for Policies and Governance to Achieve More Sustainable and Healthy Food Systems;

	References; 3 Demographic Change, Agriculture, and Rural Poverty; 3.1 Introduction; 3.2 Global Trends and Projections Population Growth and Urbanization economic Growth and Poverty; Youth Bulges and Aging Farmers; 3.3 Growth and Structural Change; Exiting Agriculture; Structural Change and Poverty Reduction; 3.4 Urbanization and Agriculture; 3.5 Conclusions; References; 4 Climate Change, Agriculture and Food Security: Impacts and the Potential for Adaptation and Mitigation; 4.1 Climate Change Challenges; 4.2 Climate Change Impacts on Agriculture and Food Security; How We Estimate Impacts; Impacts of Climate Change on Agriculture and Food Security; 4.3 Adaptation Options; Diversification Sustainable Intensification carbon and Nitrogen Management Options; Identifying and Addressing Barriers to Adoption; 4.4 Mitigation Options; Managing Carbon-Rich Forest Landscapes; Livestock Emissions; Flooded Rice; Nitrous-Oxide Emissions; Soil Carbon Management; Higher Production Efficiency, Lower Emissions, and Enhanced Food Security; Investing in Yield Improvements; Efficiency Improvements in Aquaculture and Fisheries; Mitigation Costs, Incentives, and Barriers; Beyond the Farmgate: A Food System Perspective on Emissions
Subjects	Food security.
	Food supply.
	Sustainable agriculture.
	Food crops.

Bibliography

Notes	Includes bibliographical references and indexes.

That sugar book: the essential companion to the feature documentary that will change the way you think about 'healthy' food	
LCCN	2015297300
Type of material	Book
Personal name	Gameau, Damon, author.
Main title	That sugar book: the essential companion to the feature documentary that will change the way you think about 'healthy' food / Damon Gameau.
Edition	First U.S. edition.
Published/Produced	New York: Flatiron Books, 2015.
Description	239 pages: illustrations (chiefly color); 26 cm
Links	Contributor biographical information http://www.loc.gov/catdir/enhancements/fy1607/2015297300-b.html
	Publisher description http://www.loc.gov/catdir/enhancements/fy1607/2015297300-d.html
ISBN	125008234X
	9781250082343
LC classification	QP702.S85 G36 2015
Subjects	Sugar--Health aspects.
	Sugar-free diet.
	Sugar-free diet--Recipes.
	Food--Sugar content.
Notes	Includes bibliographical references (page 235) and index.

The Blender Girl: super-easy, super-healthy meals, snacks, desserts, and drinks	
LCCN	2013046335

Type of material	Book
Personal name	Masters, Tess.
Main title	The Blender Girl: super-easy, super-healthy meals, snacks, desserts, and drinks / Tess Masters; photography by Anson Smart.
Published/Produced	Berkeley: Ten Speed Press, [2014]
Description	217 pages: color illustrations; 24 cm
ISBN	9781607746430 (paperback)
LC classification	TX840.B5 M369 2014
Scope and content	"The debut cookbook from the powerhouse blogger behind theblendergirl.com, featuring 100 gluten-free, vegan recipes for smoothies, meals, and more made quickly and easily in a blender. What's your perfect blend? On her wildly popular recipe blog, Tess Masters--aka, The Blender Girl--shares easy plant-based recipes that anyone can whip up fast in a blender. Tess's lively, down-to-earth approach has attracted legions of fans looking for quick and fun ways to prepare healthy food. In The Blender Girl, Tess's much-anticipated debut cookbook, she offers 100 whole-food recipes that are gluten-free and vegan, and rely on natural flavors and sweeteners. Many are also raw and nut-, soy-, corn-, and sugar-free. Smoothies, soups, and spreads are a given in a blender cookbook, but this surprisingly versatile collection also includes appetizers, salads, and main dishes with a blended component, like Fresh Spring Rolls with Orange-Almond Sauce, Twisted Caesar Pleaser, Spicy Chickpea Burgers with Portobello Buns and Greens, and I-Love-Veggies! Bake. And even though many of Tess's smoothies and shakes taste like

	dessert--Apple Pie in a Glass, Raspberry-Lemon Cheesecake, or Tastes-Like-Ice-Cream Kale, anyone? Her actual desserts are out-of this-world good, from Chocolate-Chile Banana Spilly to Flourless Triple-Pecan Mousse Pie and Chai Rice Pudding. Best of all, every recipe can easily be adjusted to your personal taste: add an extra squeeze of this, another handful of that, or leave something out altogether--these dishes are super forgiving, so you can't mess them up. Details on the benefits of soaking, sprouting, and dehydrating; proper food combining; and eating raw, probiotic-rich, and alkaline ingredients round out this nutrient-dense guide. But you don't have to understand the science of good nutrition to run with The Blender Girl--all you need is a blender and a sense of adventure. So dust off your machine and get ready to find your perfect blend"-- Provided by publisher.
Contents	Introduction: Birth of The Blender Girl -- The Lowdown -- Love Your Blender -- Healthy Ideas To Blend In -- The Recipes -- Smoothies & Shakes -- Appetizers, Snacks, Dips & Spreads -- Salads -- Soups -- The Main Event -- Desserts -- Drinks, Juices & Tonics -- Condiments, Sauces & Creams -- Resources: Get Your Goods Here.
Subjects	Blenders (Cooking)
	Cooking (Natural foods)
	Nutrition.
	Gluten-free diet--Recipes.
	Vegan cooking.
	Cooking / Methods / Special Appliances.
	Cooking / Vegetarian & Vegan.

	Cooking / Beverages / General.
Notes	Includes bibliographical references (pages 208-211) and index.
	Includes bibliographical references and index.

The complete month of meals collection: hundreds of diabetes-friendly recipes and nearly limitless meal combinations

LCCN	2016058163
Type of material	Book
Main title	The complete month of meals collection: hundreds of diabetes-friendly recipes and nearly limitless meal combinations / American Diabetes Association.
Published/Produced	Arlington, VA: American Diabetes Association, [2017]
	©2017
Description	204 pages: color illustrations 26 cm
ISBN	9781580406628 (spiral bound in hard cover)
	1580406629 (spiral bound in hard cover)
LC classification	RC662 .C627 2017
Related names	American Diabetes Association.
Contents	What can I eat? -- Making healthy food choices -- How to use this cookbook -- Recipe cards (Breakfast, lunch, dinner) -- Breakfast -- Lunch -- Dinner -- Sides & salads -- Dressings, salsas, & sauces -- Drinks -- Desserts.
Subjects	Diabetes--Diet therapy--Recipes.
Form/Genre	Cookbooks.
Notes	Includes index.

The diabetes fast-fix slow-cooker cookbook: Fresh twists on family favorites

LCCN	2013014119

Type of material	Book
Personal name	Hughes, Nancy S.
Main title	The diabetes fast-fix slow-cooker cookbook: Fresh twists on family favorites / by Nancy S. Hughes.
Published/Produced	Alexandria: American Diabetes Association, [2014]
Description	191 pages; 23 cm
ISBN	9781580404556 (pbk.)
LC classification	RC662 .H836 2014
Summary	"As lives get more hectic and schedules grow busier, people are looking for healthy, economical, easy, "comfort on the go" recipes. It's easy to see why the popularity of slow-cookers has skyrocketed in the past few years. Slow-cooker recipes are the perfect way to prepare delicious, home-cooked meals with less hassle and less time in the kitchen. Unfortunately, many of today's slow-cooker recipes are based on old methods that produce "stewed" results. Nancy Hughes improves this approach with the addition of fresh ingredients-- introduced to the recipe at just the right time--to add pop. For example, a simple dish of chili can be dramatically improved by chopping up and setting aside some of the raw onions, peppers, and tomatoes that are already part of the dish. By sprinkling them on top with fresh lime and a dollop of sour cream at the right point in the process, it adds an unexpectedly delightful level of flavor, crunch, color, and freshness! Many of Nancy's more than 150 recipes in the book are built just like this to rejuvenate familiar dishes. By adding small twists to beloved classics, she

has created recipes that will become instant favorites at the dining table. Not only are these new recipes fresh and fabulous, but they're also incredibly healthy and guaranteed to meet the American Diabetes Association's nutritional guidelines. These guidelines ensure that the recipes will fit into anyone's diabetes meal plan and help them better manage their blood glucose levels. Each recipe features complete nutrition information and diabetic exchanges. The book features 16 pages of color photography, showing how even a simple slow-cooker recipe can provide a beautiful meal for the dining table. The interior design is in an attractive two-color format. The introductory section contains useful information on how to make a slow cooker work for you and for a healthy lifestyle. There are tips and tricks for getting the most out of your slow cooker and how to improve flavor in simple, economical ways. Here are just a few of the delightful recipes in this book: Creamy Artichoke Parmesan Dip, Hoppin' Jalapeno Stuffers, Chipotle Raspberry Pulled Chicken, Lima Bean Hummus Pile-Ups, Herbed-Parmesan Chicken Soup, Beer Bottle Chili, Smokey Bacon Beef Goulash, Southern Shrimp Creole, Pubhouse Dark Roasted Chuck Roast, Pico de Gallo Cod with Avocado, Pecan-Topped Quinoa Stack Up, Lemon-Parsley Bright Rice, Cornbread Loaf, Chunky Pear-Apricot Fruit Spread, Rustic Peach Rice Pudding, and Strawberry-Kiwi Cake in a Pot"-- Provided by publisher.

	"With this book, the reader can enjoy great tasting dishes that will bring stress levels down and healthy food intake up. None of these recipes take more than a few minutes to prepare and the slow cooker method is effortless"-- Provided by publisher.
Subjects	Diabetes--Diet therapy--Recipes.
	Electric cooking, Slow--Recipes.
	Diabetics--Nutrition.
	Cooking / Health & Healing / Diabetic & Sugar-Free
Form/Genre	Cookbooks.
Notes	Includes index.

The easy vegan cookbook: make healthy home cooking practically effortless	
LCCN	2015930605
Type of material	Book
Personal name	Hester, Kathy, author.
Main title	The easy vegan cookbook: make healthy home cooking practically effortless / Kathy Hester.
Published/Produced	Salem, MA: Page Street Publishing Co., 2015.
Description	208 pages: color illustrations; 23 cm
Links	Contributor biographical information https://www.loc.gov/catdir/enhancements/fy1617/2015930605-b.html
	Publisher description https://www.loc.gov/catdir/enhancements/fy1617/2015930605-d.html
ISBN	1624141471
	9781624141478
LC classification	TX837 .H473 2015
Summary	"Bestselling author Kathy Hester hits the bulls-eye with a brand new cookbook to solve a big

	vegan dilemma: how to make vegan food that is fast, easy and lip-smackingly delicious. The Easy Vegan Cookbook, packed with 80 recipes, is a must-have cookbook for vegans with families, busy schedules, limited budgets and hearty appetites for healthy food that simply tastes good. With recipes like Creamy Broccoli and Potato Casserole and Veggie 'PotPie' Pasta, readers will have a stockpile of quick recipes that they can count on to be delicious. Additionally, many of the recipes are gluten-free, soy-free and oil-free, for those who have other dietary restrictions or preferences. This cookbook includes chapters such as Make-Ahead Staples and Speedy Stir-Fries, as well as recipes such as Inside-Out Stuffed Pepper Stew, Creamy Cauliflower Pesto Pasta and 'Vegged-Out' Chili. No longer will weeknight meal planning be a source of stress. With The Easy Vegan Cookbook, vegans everywhere can enjoy easy, fast and family-friendly recipes for amazing food." -- Amazon.com
Subjects	Vegan cooking.
Notes	"Includes oil-free, soy-free and gluten-free options!" -- Cover
	Includes index.

The food activist handbook: big & small things you can do to help provide fresh, healthy food for your community

LCCN	2014046828
Type of material	Book
Personal name	Berlow, Ali.
Main title	The food activist handbook: big & small things

Bibliography

	you can do to help provide fresh, healthy food for your community / Ali Berlow.
Published/Produced	North Adams, MA: Storey Publishing, [2015]
	©2015
Description	320 pages: illustrations, maps; 23 cm
ISBN	9781612121802 (pbk.)
	9781603429290 (ebook)
LC classification	HV696.F6 B464 2015
Variant title	Food activist handbook: big and small things you can do to help provide fresh, healthy food for your community
Subjects	Food relief--United States.
	Slow food movement--United States.
	Nutrition--United States.
	Consumer movements--United States.
	Community activists--United States.
Notes	Includes bibliographical references and index.

The game of eating smart: nourishing recipes for peak performance inspired by MLB superstars

LCCN	2018289901
Type of material	Book
Personal name	Loria, Julie, author.
Main title	The game of eating smart: nourishing recipes for peak performance inspired by MLB superstars / Julie Loria with Chef Allen Campbell; photographs by Ben Fink.
Edition	First edition.
Published/Produced	New York: Rodale Books, an imprint of the Crown Publishing Group, [2019]
	©2019
Description	239 pages: color illustrations; 24 cm
ISBN	9781635652703 (hardcover)

	1635652707 (hardcover)
	(ebook)
LC classification	TX741 .L665 2019
Related names	Campbell, Allen, author.
	Fink, Ben, photographer.
Summary	"In Major League Baseball, the transition to eating healthy food has become more than a movement; it's a revolution. Jose Altuve, Chris Archer, Clayton Kershaw, Noah Syndergaard, Mike Trout, and the 16 other star ballplayers featured in [this book] are proof of the positive effect of proper nutrition on athletic performance and overall health. Eating smart isn't about calorie-counting and fad diets. It simply means consuming more nourishing food--including leafy greens, lean protein, and fresh fruit--that eventually decreases the desire to make unhealthy choices. [This book] includes insights from today's top players on their approach to healthy living and performance nutrition, plus more than 80 easy-to-prepare and nutrient-dense recipes inspired by their food philosophies and favorite meals" -- Provided by publisher.
Contents	Cooking tips and techniques for eating smart: Stock your kitchen; Cooking beans and grains; Kitchen essentials -- The lineup and game-changing recipes: José Altuve; Chris Archer; Nolan Arenado; Jake Arrieta; JoséBautista; Kris Bryant; Carlos Correa; Freddie Freeman; Paul Goldschmidt; Didi Gregorius; Bryce Harper; Adam Jones; Matt Kemp; Clayton Kershaw; Ian Kinsler; Corey Kluber; Hunter Pence; David Price; Ciancarlo Stanton; Noah Syndergaard;

		Mike Trout.
Subjects		Cooking (Natural foods)
		Athletes--Nutrition.
		Nutrition.
		Cooking
Form/Genre		Cookbooks.
Notes		Includes index.

The Laura Lea balanced cookbook: 120+ everyday recipes for the healthy home cook

LCCN	2019952521
Type of material	Book
Personal name	Lea, Laura, author.
Main title	The Laura Lea balanced cookbook: 120+ everyday recipes for the healthy home cook / Laura Lea.
Published/Produced	Whites Creek: Blue Hills Press, 2019.
ISBN	9781951217006 (hardback)
	(ebook)
Summary	"In Laura Lea Goldberg's new cookbook, The Laura Lea Balanced Cookbook, the rubber of old-fashioned home-cooking meets the road of new healthy-food. With over 120 approachable, comforting, make-ahead recipes, this first cookbook from the creator of the popular "LLBalanced" website reaffirms that balance is possible: you can find the joy, relaxation, and healing of cooking for yourself, family, and friends during these frenetic times. All of the recipes in are simple, familiar, and no-fuss. The majority of the recipes come together in thirty minutes or less and all are appealing to kids and adults alike, can be modified for picky eaters or

	can be proudly served at a dinner party. The food isn't dogmatic: a little of everything is used and flexibility is the key. With a focus on quality and moderation, the healthy aspects don't hit you over the head. They just make you feel good. With helpful shopping lists and easy-to-follow menu plans, The Laura Lea Balanced Cookbook will help any home cook create a foundation in the pantry and kitchen that will make the prospect of healthy cooking accessible and exciting, not stressful. It doesn't overthink things and focuses on consistency instead of perfection. In the end, The Laura Lea Balanced Cookbook will have you discovering the balance of cooking delicious, healthy meals at home while re-connecting with yourself, family, and friends"-- Provided by publisher.
The nation's favourite healthy food	
LCCN	2016429314
Type of material	Book
Personal name	Maguire, Neven, author.
Main title	The nation's favourite healthy food / Neven Maguire.
Published/Produced	Dublin: Gill & Macmillan, [2015]
Description	18 unnumbered pages, 252 pages: color illustrations; 25 cm
ISBN	9780717167999 (hbk.)
	0717167992 (hbk.)
LC classification	TX741 .M336 2015
Subjects	Cooking (Natural foods)
	Reducing diets--Recipes.
Form/Genre	Cookbooks.

The new American Heart Association cookbook	
LCCN	2017288949
Type of material	Book
Main title	The new American Heart Association cookbook / American Heart Association.
Edition	9th edition.
Published/Produced	New York: Harmony Books, [2017]
	©2017
Description	535 pages; 24 cm
ISBN	9780553447187 (hardcover)
	0553447181 (hardcover)
	9780553447200 (paperback)
	(ebook)
LC classification	RC684.D5 A44 2017
Related names	American Heart Association, editor.
Summary	Here is the ultimate resource for anyone looking to improve cardiac health and lose weight, offering 800 recipes--100 all new, 150 refreshed--that cut saturated fat and cholesterol. The American Heart Association's cornerstone cookbook has sold more than three million copies and it's now fully updated and expanded to reflect the association's latest guidelines as well as current tastes, with a fresh focus on quick and easy. This invaluable, one-stop-shopping resource -- including updated heart-health information, strategies and tips for meal planning, shopping, and cooking healthfully -- by the most recognized and respected name in heart health is certain to become a staple in American kitchens.-- Source other than Library of congress.
Contents	Eat well to stay well. Making healthy food

	choices; Making healthy lifestyle choices; About the recipes -- Recipes. Appetizers, snacks, and beverages; Soups; Salads and salad dressings; Seafood; Poultry; Meats; Vegetarian entrées; Vegetables and side dishes; Sauces and gravies; Breads and breakfast dishes; Desserts -- Appendixes. How your diet affects your heart; Shopping with your heart in mind; Cooking for a healthy heart; Menu planning for holidays and special occasions; Equivalents and substitutions.
Subjects	Heart--Diseases--Diet therapy--Recipes.
	Low-cholesterol diet--Recipes.
	Heart--Diseases--Diet therapy--Recipes.
	Cooking / Health & Healing / Heart.
	Cooking / Health & Healing / Low Fat.
	Heart--Diseases--Diet therapy.
	Low-cholesterol diet.
Form/Genre	Cookbooks.
	Recipes.
Notes	Includes index.
	"Revised and updated with more than 100 all-new recipes"--Dust jacket.

The wicked healthy cookbook: free. from animals.

LCCN	2018288838
Type of material	Book
Personal name	Sarno, Chad, author.
Main title	The wicked healthy cookbook: free. from animals. / Chad Sarno, Derek Sarno, and David Joachim; foreword by Woody Harrelson; photographs by Eva Kosmas Flores.
Edition	First edition.
Published/Produced	New York: Grand Central Life & Style, 2018.

Description	xii, 306 pages: color illustrations; 26 cm
ISBN	9781455570287 (hardcover)
	1455570281 (hardcover)
	(ebook)
LC classification	TX837 .S26536 2018
Related names	Sarno, Derek, author.
	Joachim, David, author.
	Harrelson, Woody, writer of foreword.
	Flores, Eva Kosmas, photographer.
Summary	"[This book] takes ... plant-based cooking to a whole new level. The chefs have pioneered innovative cooking techniques such as pressing and searing mushrooms until they reach a rich and delicious meat-like consistency. Inside, you'll find informative sidebars and must-have tips on everything from oil-free and gluten-free cooking (if you're into that) to organizing an efficient kitchen"--Amazon.com.
Contents	Time to get wicked healthy -- What to keep on hand -- The conscious cook's mind-set -- Healthy food doesn't have to taste like shit -- First bites: Nibbles; Toasts; Dips and spreads; Plated and shareable appetizers -- Handhelds: Pizza; Tacos; Burgers and sandwiches -- Bowls: Vegetable, bean, and grain bowls; Soups and stews; Oatmeal bowls -- Straight up vegetables: Straight up vegetables; Salads -- Comfort food: Pasta and risotto; Food bars and shareable platters -- Nature's candy: Fruit desserts; Cookies and bars; Cakes, tarts, and puddings -- Wicked healthy juices and cocktails: Juices and coolers; Cocktails -- Sauces and basics: Sauces, dressings, butters, and glazes; Salsas and hot

	sauces; Pickles and preserves; Stocks and broths -- Appendix: Wicked healthy benders and special diets.
Subjects	Vegan cooking.
	Cooking (Natural foods)
Form/Genre	Cookbooks.
Notes	Includes index.

Twenty years of life: why the poor die earlier and how to challenge inequity	
LCCN	2018933658
Type of material	Book
Personal name	Bohan, Suzanne, author.
Main title	Twenty years of life: why the poor die earlier and how to challenge inequity / Suzanne Bohan.
Published/Produced	Washington, DC: Island Press, [2018]
	©2018
Description	253 pages, 8 unnumbered pages of plates: color illustrations; 24 cm
ISBN	9781610918015 (hbk.)
	1610918010 (hbk.)
LC classification	HC79.P6 B582 2018
Portion of title	Why the poor die earlier and how to challenge inequity
Other title	20 years of life: why the poor die earlier and how to challenge inequity
Related names	Island Press, publisher.
Summary	"In Twenty Years of Life, Suzanne Bohan exposes the flip side of the American dream: your health is largely determined by your zip code. The strain of living in a poor neighborhood, with subpar schools, lack of parks, fear of violence, and few to no healthy

	food options is literally taking years off people's lives. The difference in life expectancy between rich and poor neighborhoods can be as much as twenty years. In a bold experiment to challenge this inequity, the California Endowment is upending the top-down charity model by investing 1 billion dollars over ten years to help distressed communities advocate for their own interests. The key is unleashing the political power of residents, who are pushing reform both locally and in the state's legislative chambers. If it works in fourteen of California's most challenging and diverse communities, it can work anywhere in the country. In this revealing and inspiring book, Bohan tells the stories of former convicts who now work to prevent gun violence; kids who convinced their city council to build skate parks; and students who demanded fairer school discipline policies. We meet urban farmers who fought for the right to sell their produce and a Native American tribe that is restoring its health by first restoring its ancestral land. Told with compassion and insight, their stories will fundamentally change how we think about the root causes of disease and the prospects for healing"-- Provided by publisher.
Contents	How neighborhoods kill -- The stress effect -- Keeping kids in school -- Changing schools' rules -- A safe place to play -- A safe place to live -- Rural activism -- Good eats - Healing trauma -- Red and blue visions of health -- Epilogue: 209.

Subjects	California Endowment.
	2000-2099
	Poverty--United States--Social conditions--21st century.
	Equality--California.
	Poor--California.
	Social classes--California.
	Violence--California.
	Human rights--California.
	Educational equalization--California.
	Food security--California.
	Social Science / Disease & Health Issues.
	Social Science / Human Services.
	Social Science / Social Classes & Economic Disparity.
	Economic history.
	Educational equalization.
	Equality.
	Food security.
	Human rights.
	Poor.
	Social classes.
	Violence.
	United States--Economic conditions.
	California.
Notes	Includes bibliographical references (pages 221-245) and index.

Yum, yum, baby!: first words for little foodies	
LCCN	2018295921
Type of material	Book
Personal name	Wren, Rosalee, author.
Main title	Yum, yum, baby!: first words for little foodies /

	written by Rosalee Wren, illustrated by Kat Uno.
Published/Produced	Barrington, Illinois: Cottage Door Press, [2018]
Description	1 volume (unpaged): color illustrations; 19 cm
ISBN	9781680522778 (hbk.)
	1680522779
LC classification	PZ7.1.W73 Yu 2018
Related names	Uno, Kat, illustrator.
Summary	Charming, colorful illustrations introduce first words for little foodies! This padded board book encourages healthy food choices and supports comprehension and vocabulary growth.
Subjects	Board books.
	Vocabulary--Juvenile fiction.
	Food--Juvenile fiction.
	Nutrition--Juvenile fiction.
	Vocabulary--Fiction.
	Food--Fiction.
	Nutrition--Fiction.
	Board books.
	Food.
	Nutrition.
	Vocabulary.
Form/Genre	Board books.
	Fiction.
	Juvenile works.
	Board books.
Notes	12m+

RELATED NOVA PUBLICATIONS

COST REALITIES OF HEALTHY FOODS

Editors: Regina Snyder and Robert P. Wagner

ISBN: 978-1-62257-350-9
Series: Food and Beverage Consumption and Health and Food Science and Technology
Binding: Hardcover
Publication Date: October 2012

Most Americans consume diets that do not meet Federal dietary recommendations. This perception may be influenced by studies that found healthy foods to cost more per calorie than less healthy foods. This is one way, but not the only way, to measure the cost of a healthy diet.

For a balanced assessment, this book compares the price of healthy and less healthy foods using three metrics: the price per calorie, per edible gram, and per average portion. These studies conclude that the higher prices of healthy foods present barriers to consumer ability to buy recommended amounts of foods like fruits and vegetables

"Healthy Food," "Earthing," and the Myth of the Dangerous Free Radicals[*]

Søren Ventegodt[1,2,3,4,5,†]

[1]Quality of Life Research Center, Copenhagen, Denmark
[2]Research Clinic for Holistic Medicine
[3]Nordic School of Holistic Medicine, Copenhagen, Denmark
[4]Scandinavian Foundation for Holistic Medicine, Sandvika, Norway
[5]Interuniversity College, Graz, Austria

The world is getting crazy. In what should be an awakening towards truth about medicine's old obsolete ideas are left and substituted with even older and more obsolete ideas. One of the oldest and most incorrect ideas in biochemistry was the myth about "dangerous free radicals" causing everything evil from age to cancer presented about 1956 by Denham Harman (1, 2).

---

[*] The full version of this chapter can be found in *Alternative Medicine Research Yearbook 2015*, edited by Joav Merrick, MD, published by Nova Science Publishers, Inc, New York, 2015.
[†] Corresponding author: Søren Ventegodt, MD, MMedSci, EU-MSc-CAM, Director, Quality of Life Research Center, Frederiksberg Alle 13A, 2tv, DK-1661 Copenhagen V, Denmark. E-mail: ventegodt@livskvalitet.org.

Later it was discovered that both the production and the destruction of all the reactive oxygen species (ROS) and nitrogen-based free radicals are carefully controlled by the cells enzyme system, and that free radicals are used in the living organism for a number of essential and often life-saving purposes like destruction of bacteria in the phagocyte and regulation of smooth muscle tonus i.e., in the intestine (3).

In the lab free radicals, or radicals, are highly reactive molecules, so explosive and dangerous that even the modern chemist cannot always control them. In the body they are simply under total biological control.

And more than that. The enzyme catalase and other enzymes related to the metabolism of free radicals exists in almost all organisms on earth, telling us that oxygen and free radicals was a big problem early in evolution 3-4 billion years ago, and that it was solved effectively, and that the solution was so good that it stayed with living organism for eternity. Catalase is even the fastest working enzyme known to man, indicating that life really did a good job solving this problem.

If we add active oxygen to cells (called "oxygen stress") in the lab we can see that then the catabolism is simply raised to counterbalance the higher concentration of oxygen and free radicals.

So we can safely conclude that life is amazing in its control of its molecules, and that free radicals are not the slightest problem for normal, healthy modern cells. People who believe in the mortal dangers of free radicals simply do not understand the normal biochemistry of the cell.

But the old myth still lives with many people who even seem to benefit from this belief. Experts in nutrition uses free radicals as argument for our need for fruit and vegetables, and countless "alternative therapists" are treating people with all kind of diets and drugs and vitamins that are supposed to "work against the free radicals" because of their antioxidant properties.

FOOD AND HEALTHY MIND: CAFFEIC ACID AND ITS DERIVATIVES[*]

Satyabrata Ghosh
Department of Ingeniería Rural, ETSI Agrónomos,
Universidad Politécnica de Madrid (UPM), Madrid, Spain

With rapid socio-ecomonic advancement and compulsion for cosmopolitan human society, the pressure of overwork is bending the society to its toe. Therefore, the result is different neurological and neurovascular diseases, cardiac diseases and other brain diseases are growingly increased day by day. Human food habits are strikingly different from one country to other and even with the difference of their religions, culture, society structures and economic conditions. Indian "Vedas" is first text to describe the deep rooted relation between foods & mind. The type of foods and therefore its importance for health have an interim connection with the change of climate and play a great role to build mind and health of the individual and the society as a whole. Upanishads tell that 'Annam Brahma' (i.e., Food is God) and state the quality of the mind is influenced by the quality of the food intake. So in this above context this article tries to critically describe an eternal relation of food and healthy mind in terms of antioxidant property. Here through this article we mainly want to highlight the role of caffeic acid because caffeic acid is a globally available antioxidant compound and since the last more than a decade researchers have found that caffeic acid and its derivatives have lots of effects on human mind and thus on health. Neurological studies have shown that caffeic acid can create protective effects on neurodegeneration,

---

[*] The full version of this chapter can be found in *Caffeic Acid: Biological Properties, Structure and Health Effects*, edited by Leanna Vaughn., published by Nova Science Publishers, Inc, New York, 2015.

neurotoxicity and has potential cytoprotective effects against a wide range of injuries in different tissues, including the brain.

NUTRITION EDUCATION AND THE COST OF HEALTHY FOOD – DO THEY COLLIDE? LESSONS LEARNED IN A PREDOMINANTLY BLACK URBAN TOWNSHIP IN SOUTH AFRICA[*]

Moïse Muzigaba[†], MPH, and Thandi Puoane, DrPH

Faculty of Community and Health Sciences,
School of Public Health, University of the Western Cape, Belleville,
South Africa and The Heart and Stroke Foundation, South Africa

The cost of healthier foods has been shown to contribute negatively to individuals' food choices in developed societies. However, there is a dearth of knowledge regarding this phenomenon in low to middle income countries, particularly in Africa. This study explored community members' experiences in buying healthier food options and compared their perceived cost of selected healthier and less healthy foods with actual market costs. The study was conducted amongst 50 adult health club members in Khayelitsha in the Western Cape Province of South Africa, using both quantitative and qualitative research methods. Data were gathered in three phases: The first phase involved interviews with all the 50 participants.

---

[*] The full version of this chapter can be found in *Public Health Yearbook 2013*, edited by Joav Merrick, MD, MMedSci, DMSc, published by Nova Science Publishers, Inc, New York, 2014.
[†] Correspondence: Moïse Muzigaba, Faculty of Community and Health Sciences, School of Public Health, University of the Western Cape, Private Bag X17 Bellville, 7535, South Africa. E-mail: mochemoseo@gmail.com and The Heart and Stroke foundation South Africa, POBox 15139 Vlaeberg 8018, Cape Town, South Africa. E-mail: moise@heartfoundation.co.za.

The second phase involved in-depth interviews with ten purposively selected members. In the third phase, food price audits were conducted in supermarkets and convenient stores in the study setting. Quantitative data were subjected to descriptive statistical analysis, while content analysis was used to analyze qualitative data. Our quantitative findings showed that most of the members were illiterate, unemployed, and largely dependent on government grants. Qualitative findings showed that low household incomes, inability to read and interpret nutritional information and personal food preferences contributed to community members' unhealthy food-purchasing behavior. From both local store audits and participants' perceptions, healthier foods tended to be more expensive than their less healthy options. Low income was a major factor militating against participants' healthy food-purchasing behavior. Future research studies are needed to assess how trends in food prices over time affect individuals' healthy food purchasing behaviors.

EUROPEAN CONSUMERS ACCEPTANCE OF HEALTHY FOOD PRODUCTS: A REVIEW OF FUNCTIONAL FOODS[*]

Azzurra Annunziata[1,†] and Riccardo Vecchio[2]

[1]University of Naples "Parthenope", Naples, Italy
[2]University of Naples Federico II, Portici, Italy

In the recent past, Europe has experienced an increase in several chronic diseases linked to dietary and lifestyle habits.

---

[*] The full version of this chapter can be found in *Functional Foods: Sources, Biotechnology Applications and Health Challenges*, edited by Aron Robinson and Dominick Emerson, published by Nova Science Publishers, Inc, New York, 2012.
[†] azzurra.annunziata@uniparthenope.it.

In particular, obesity is continuously increasing at an alarming rate all over Europe.

As a result healthier food products have rapidly gained important market shares. The food industry has rapidly reacted to this trend by developing a growing variety of new products with health-related claims and images, including functional foods (FF) that are selected by consumers for their health-promoting properties. However, demand for FF within the European Union varies considerably from country to country, mainly due to food traditions and different cultural heritage. This chapter offers an overview of the current FF market situation in Europe and addresses the main cognitive, motivational and attitudinal determinants of consumer acceptance of these foods in different European countries, providing practical insights of the profile of FF consumers. In particular, the major factors that are currently restraining growth of FF in Europe (beyond economic crisis) are highlighted and deeply discussed. As the lack of categorisation (it is still difficult for consumers to distinguish between functional and conventional food products); growing consumer confusion (due to increasing product variety and product labelling, together with the overwhelming amount of general marketing and educational messages on food intake); low awareness of health benefits of functional ingredients; lack of consensus on the level of scientific support and documentation required for specific health claims. The current work could help practitioners in making more effective strategic and tactical marketing decisions, and add some useful information for government bodies interested in designing public health programs.

MEAT PRODUCTS AS HEALTHY FOOD: USE OF VEGETABLE OILS AND OTHER FUNCTIONAL INGREDIENTS IN THEIR FORMULATION[*]

Alfonso Totosaus[1,] and M. Lourdes Pérez-Chabela[2#]*

[1]Food Science Lab, Tecnológico de Estudios Superiores de Ecatepec. Estado de Mexico, Mexico
[2]Departamento de Biotecnología, Universidad Autónoma Metropolitana Iztapalapa, Iztapalapa, Mexico City, Mexico

Meat is the main source of animal protein for human, contributing to essential amino acids and other important nutrients, like iron. Nonetheless, and although in developed countries red meat consumption has been recently associated to some health concerns, in emerging economies of developing countries the demand for red meat has been increased. The fact is that the price of red meat in some countries made that their consumption being low. Processed meat products offers a wide range of products at lower price that can be consumed for most of the population. However, fat and salt in the formulation could be associated as well to health issues. Nutritional properties of emulsified meat products can be enhanced by many options, from reducing salt to replacing animal fat with vegetable oils rich in unsaturated fatty acids. The incorporation of dietetic fiber as prebiotic is also a possibility, where their use jointly with thermotolerant probiotic lactic acid bacteria could be a viable alternative to symbiotic low fat salt reduced healthy meat products.

---

[*] The full version of this chapter can be found in *Meat Consumption and Health*, edited by Maria Pilar Ortega and Rafael Soto, published by Nova Science Publishers, Inc, New York, 2012.

[*] Corresponding author: Alfonso Totosaus. Food Science Lab, Tecnológico de Estudios Superiores de Ecatepec. Av. Tecnológico y Av. Central, Ecatepec CP 55210, Estado de Mexico, Mexico. E-mail: atotosaus@tese.edu.mx.

[#] M. Lourdes Pérez-Chabela. Departamento de Biotecnología, Universidad Autónoma Metropolitana Iztapalapa, Av. San Rafael Atlixco 186, Iztapalapa CP 09270, Mexico City, Mexico.

INDEX

A

acid, 25, 31, 42, 43, 59, 64, 66, 73, 76, 92, 102, 103, 162
active compound, 41
active oxygen, 161
adolescents, 23, 27, 30, 31, 32, 34, 35, 37, 38
adulthood, viii, 22, 25, 26, 27, 28, 29, 30, 36
age, viii, 7, 8, 9, 22, 35, 53, 67, 160
agricultural sector, 136
agriculture, 85, 105, 135, 138
agro industrial coproducts, 40
anorexia, 3, 5, 8, 10, 11, 12, 13, 18, 123
anorexia nervosa, 5, 14, 18
antigen-presenting cells, 25
antioxidant, ix, 29, 40, 43, 48, 49, 56, 57, 73, 119, 120, 121, 161, 162
antioxidant capacity, ix, 40, 42, 48, 49, 56, 57, 120
antioxidants, ix, 31, 40, 41, 46, 55, 57, 71, 104, 105, 121
anxiety, 3, 5, 10, 11, 14, 17
anxiety disorder, 10, 11
attachment, 10, 14, 15, 18

attitudes, 4, 8, 10, 11, 17, 60
authenticity, 122, 126, 127
autoimmune disease, 24, 25
awareness, 11, 14, 40, 165

B

bacteria, 40, 43, 46, 49, 50, 51, 53, 92, 161, 166
bacterial fermentation, 25
behaviors, 7, 8, 11, 22, 23, 24, 28, 33, 112, 164
beneficial effect, 9, 23, 24, 31, 48
benefits, 55, 75, 107, 108, 141, 165
beverages, 27, 30, 60, 115, 133, 152
binding globulin, 28
binge eating disorder, 8
biochemistry, 104, 160, 161
bio-indicators, 122
biological control, 161
biomarkers, 37
blood, 3, 24, 27, 46, 63, 86, 144
blood pressure, 24, 27
body fat, 38
body image, 4, 5, 10, 15

body mass index, 19
body shape, 10
body weight, 23, 24, 64
bonds, 45, 65, 97
bone mass, 30, 31, 37
bone mineral content, 38
breast cancer, 28, 29
breastfeeding, 34
bulimia, 3, 8, 10, 12, 14, 18
bulimia nervosa, 3, 14, 18
by-products, 57, 59

C

calorie, 26, 148, 159
cancer, 28, 29, 36, 37, 78, 160
cancer cells, 29
cancer death, 28
candelilla wax, v, vii, x, 61, 62, 64, 76
carbohydrates, 24, 25, 42, 43, 48, 49
carbon, 43, 138
carcinogenesis, 37
cardiovascular disease, viii, ix, 22, 26, 35, 61, 63
cardiovascular diseases, 63
cardiovascular risk, 26
carotenoids, 29, 49, 58
case studies, 5, 102, 110
cellulose, 45, 65, 72
cellulose derivatives, 72
chemical, 48, 56, 57, 72, 104
chicken, 56, 74, 107, 108
childhood, vii, viii, ix, 3, 22, 23, 24, 25, 26, 27, 28, 29, 30, 31, 32, 34, 36, 37, 83
children, viii, 22, 23, 24, 25, 26, 27, 28, 30, 31, 32, 33, 34, 35, 37, 78, 79, 83, 101, 112, 124, 131
cholesterol, 26, 27, 46, 48, 63, 151, 152
chronic diseases, ix, 22, 164
chronic illness, 13

classification, 77, 79, 80, 81, 82, 83, 84, 85, 86, 88, 89, 90, 91, 94, 96, 97, 98, 100, 101, 103, 104, 105, 106, 107, 109, 110, 112, 113, 114, 115, 117, 118, 120, 122, 123, 124, 125, 126, 127, 128, 130, 131, 132, 135, 136, 139, 140, 142, 143, 145, 147, 148, 150, 151, 153, 154, 157
climate change, 134, 136
color, iv, vii, ix, x, 39, 41, 44, 46, 50, 51, 53, 55, 58, 62, 64, 66, 67, 70, 71, 72, 79, 81, 83, 84, 93, 106, 107, 108, 109, 115, 116, 118, 124, 130, 131, 135, 139, 140, 142, 143, 145, 147, 150, 153, 154, 157
colorectal cancer, 29, 92
communities, 9, 13, 84, 93, 126, 134, 136, 155
composition, ix, 40, 42, 48, 49, 57, 64, 72, 85
compounds, ix, 29, 31, 41, 42, 48, 53, 55, 57, 61, 104
consumers, ix, 40, 41, 53, 55, 71, 129, 165
consumption, 11, 23, 24, 26, 27, 29, 30, 32, 33, 34, 63, 95, 101, 129, 166
contamination, 50
content analysis, 164
control group, 7, 8
controversial, 9
controversies, 102
cooking, 45, 46, 53, 80, 81, 82, 83, 106, 107, 108, 116, 117, 130, 131, 141, 145, 146, 149, 151, 153, 154
cooperation, 37, 38, 129
coronary heart disease, 26, 34, 63, 64
cross-sectional study, 23, 34
cultural heritage, 165
cultural values, 12
culture, vii, 1, 43, 48, 59, 102, 126, 127, 129, 162
culture medium, 43

D

deficiency, viii, 3, 22
deformation, 45, 65
democracy, 102, 131
depressive symptomatology, 18
developed countries, 166
diabetes, viii, 22, 24, 25, 34, 64, 86, 142, 143, 144
diagnostic criteria, 4, 17
diastolic blood pressure, 27
diet, viii, 4, 5, 8, 9, 22, 23, 27, 29, 30, 31, 32, 34, 35, 36, 38, 63, 64, 77, 78, 87, 88, 89, 90, 91, 92, 95, 100, 102, 107, 108, 109, 118, 119, 139, 141, 152, 159
dietary fat, 27, 64
dietary fiber, ix, 24, 27, 40, 42, 46, 56, 74
dietary habits, 3, 13, 23, 24, 26, 34, 35
dietary intake, 33
disease progression, 32
diseases, vii, viii, 5, 12, 13, 22, 27, 32, 40, 162
disorder, 2, 8, 10, 12, 13, 16, 19, 20
distilled water, 43, 66
distress, 4
dressings, 78, 117, 152, 153
dyslipidemia, 26, 27

E

eating disorder, viii, 1, 3, 7, 8, 9, 10, 12, 14, 16, 18, 19, 22, 123
e-commerce, 130
economic crisis, 165
economic status, 23
education, 19, 23, 24, 88, 89, 91, 124
electrochemistry, 121
emerging markets, 129
emotional intelligence, 14
energy, 26, 29, 64, 65, 109, 110, 119
energy density, 26
energy efficiency, 110
energy expenditure, 64
environment, viii, 1, 3, 13, 32, 96, 105, 111
environmental sustainability, 136
epidemiology, 35, 37
evidence, viii, 22, 25, 36, 99
executive functions, 11
experimental condition, 71

F

families, 23, 112, 146
family members, 23
farmers, 84, 105, 155
farmland, 105
farms, 105, 132
fast food, 23, 30, 97, 98
fat, vii, x, 24, 26, 27, 29, 30, 33, 35, 41, 48, 54, 55, 62, 63, 64, 65, 67, 68, 69, 71, 72, 73, 74, 75, 76, 166
fat intake, 26
fat replacement, vii, 62
fatty acids, 24, 26, 29, 32, 34, 35, 63, 72, 73, 76, 166
feeding behavior, 22
fermentation, 43, 56
fiber, ix, 24, 27, 29, 40, 41, 42, 46, 47, 48, 49, 51, 53, 54, 55, 56, 59, 74, 166
fiber content, 41, 48
fibers, 53, 92
fish, 27, 30, 31, 115
fitness, 17, 35, 38
flavor, vii, ix, 39, 47, 50, 53, 54, 63, 78, 117, 143
flour, vii, ix, 39, 42, 43, 44, 45, 47, 48, 49, 50, 51, 52, 53, 54, 55, 59, 107, 108
food, viii, ix, 2, 3, 5, 8, 9, 10, 11, 12, 15, 17, 22, 23, 24, 28, 29, 31, 32, 33, 37, 41, 42, 46, 53, 59, 60, 61, 63, 73, 75, 77, 78, 79, 80, 81, 82, 83, 84, 85, 86, 87, 88, 89, 90,

91, 93, 94, 96, 97, 99, 100, 101, 103, 104, 105, 106, 107, 108, 109, 110, 112, 113, 114, 115, 116, 118, 120, 122, 123, 124, 126, 127, 129, 131, 132, 135, 136, 139, 140, 142, 145, 146, 147, 148, 149, 150, 151, 153, 155, 156, 157, 162, 163, 165
food habits, viii, 22, 162
food industry, 73, 165
food intake, viii, 22, 23, 145, 162, 165
food neophobia, v, 39, 41, 46, 53, 54, 59
food production, 104
food products, ix, 31, 61, 63, 91, 120, 165
food security, 136
free radicals, 160, 161
fructose, 27, 31
fruits, 23, 26, 28, 30, 31, 32, 48, 57, 160
functional food, 53, 55, 56, 58, 121, 165
functional ingredients, v, vii, ix, 39, 40, 55, 56, 165, 166

G

gastrointestinal tract, 49
gender differences, 7
gene expression, 29
gene promoter, 29
generalized anxiety disorder, 10
glucose, 43, 46, 48, 86, 144
growth, viii, 22, 26, 27, 28, 29, 30, 40, 43, 46, 50, 53, 136, 157, 165
growth factor, 28
growth rate, 136
growth spurt, 30
guidelines, 35, 87, 88, 89, 90, 101, 119, 144, 151

H

health, vii, viii, ix, x, 3, 4, 5, 7, 8, 11, 14, 22, 27, 30, 31, 33, 34, 35, 37, 38, 40, 46, 57, 62, 63, 64, 73, 78, 86, 87, 88, 89, 90, 91, 92, 98, 100, 101, 103, 104, 109, 113, 134, 148, 151, 154, 155, 162, 163, 165, 166
health care, 35
health effects, 3, 27, 46
health information, 87, 88, 89, 90, 151
health problems, 92
healthy, v, vii, viii, ix, 1, 2, 3, 4, 5, 6, 7, 8, 9, 10, 11, 12, 13, 15, 16, 17, 18, 20, 21, 22, 23, 24, 26, 28, 30, 33, 36, 39, 40, 55, 61, 77, 79, 80, 81, 82, 83, 84, 86, 87, 88, 89, 90, 91, 92, 94, 95, 96, 97, 99, 100, 101, 103, 104, 105, 106, 107, 108, 109, 110, 111, 112, 113, 114, 115, 116, 118, 120, 122, 123, 124, 126, 127, 128, 129, 130, 131, 134, 135, 137, 139, 140, 141, 142, 143, 145, 146, 147, 148, 149, 150, 151, 152, 153, 154, 157, 159, 160, 161, 162, 163, 164, 166
healthy eating, vii, 1, 2, 4, 6, 7, 8, 10, 11, 12, 13, 16, 18, 77, 82, 94, 110, 111, 112, 116, 123, 135
healthy eating disorder, 2
healthy eating pathology, 2
healthy food dependence, 2
healthy meat products, v, 39, 40, 166
heart disease, 63
heart rate, 3
heating rate, 74
histone deacetylase, 25
human, 56, 91, 92, 117, 162, 166
human body, 92
human development, 91

I

incidence, ix, 22, 32
income, 24, 33, 163
individuals, 13, 24, 163
inequity, 154, 155
infancy, 24, 27, 31, 35
ingredients, vii, ix, 2, 39, 40, 41, 48, 54, 55, 56, 73, 79, 117, 141, 143, 165
intervention, 7, 23, 24, 25, 26, 31, 32, 33, 35

L

lactic acid, 40, 49, 50, 51, 53, 166
leukemia, 28, 29, 36
life expectancy, 155
lifestyle behaviors, 28
lifetime, 109, 111, 112
light, 6, 7, 8, 10, 14, 71, 117, 134
lipid metabolism, 28
lipid oxidation, 49, 57, 64, 71
low-density lipoprotein, 26, 63
luminosity, 44, 50, 51, 71

M

malnutrition, 136
management, 86, 111, 130
marketing, 73, 101, 104, 105, 128, 129, 130, 165
matrix, 30, 53, 68, 73
matter, iv, 113, 119
measurement, 6, 12, 20, 121
meat, vii, ix, 28, 29, 30, 31, 32, 40, 41, 45, 46, 47, 48, 49, 50, 53, 55, 56, 57, 58, 59, 61, 63, 64, 65, 66, 68, 71, 72, 73, 74, 75, 78, 79, 102, 115, 117, 127, 128, 153, 166
medical, 3, 8, 9, 15, 17, 87, 88, 89, 90
medical reason, 9
medicine, 15, 92, 100, 160

mental disorder, 13
mental illness, 11, 12
meta-analysis, 15, 30, 33
metabolic acidosis, 3
metabolic syndrome, 34
metabolism, 31, 35, 48, 161
microbiota, 25, 48, 92
micronutrients, 48
microorganisms, 50
moisture, vii, ix, x, 40, 41, 42, 45, 50, 51, 52, 53, 58, 62, 65, 67, 68, 69, 75
moisture content, 68
molecules, 117, 120, 161
mortality, 28, 29, 36
multidimensional, 10, 13

N

natural resources, 136
neophobia, ix, 40, 46, 53, 56
neurodegeneration, 162
non-communicable diseases, v, vii, viii, 21, 22, 32
nutraceutical, 120
nutrient, viii, 22, 26, 31, 46, 87, 88, 89, 90, 141, 148
nutrition, 3, 8, 9, 13, 23, 24, 30, 31, 32, 33, 34, 35, 36, 37, 38, 56, 58, 86, 87, 88, 89, 90, 91, 92, 109, 112, 141, 144, 148, 161

O

obesity, ix, 9, 26, 27, 28, 33, 34, 36, 40, 61, 92, 102, 165
obsessive-compulsive disorder, 4, 16
oil, 62, 63, 64, 71, 72, 74, 76, 146, 153
oleogel, vii, x, 61, 62, 64, 65, 67, 68, 69, 70, 71, 72, 73, 75
olive oil, 27, 31, 75
organogels, 62, 74, 75, 76

Index

orthorexia nervosa, v, vii, viii, 1, 2, 4, 13, 14, 15, 16, 17, 18, 19, 20
overweight, 10, 28, 33, 40
oxidative rancidity, x, 41, 62, 66, 67, 70, 71
oxidative stress, 29

P

parental control, viii, 22
parental support, 23, 33
parenting, viii, 22, 23, 24, 33
parenting styles, 23
parents, viii, 22, 23, 24, 79, 97, 112
participants, 6, 9, 10, 17, 67, 163
pasta, 107, 108, 115, 116
perfectionism, 10, 11, 15
peripheral blood, 25
personality, 1, 10, 12, 13, 17, 18
personality characteristics, 12
phenolic compounds, 41, 58
photographs, 107, 108, 124, 147, 152
physical activity, 7, 23, 32, 33, 37, 64
physical characteristics, 72
physical fitness, 12
physical properties, 64
polyphenols, ix, 29, 40, 42, 48, 49
polyunsaturated fat, 26, 63, 75
polyunsaturated fatty acids, 26, 63, 75
population, 6, 7, 8, 9, 16, 19, 31, 36, 166
prebiotic, ix, 40, 41, 43, 48, 49, 50, 51, 53, 54, 166
prebiotic effect, 41, 53
prebiotic ingredients, 54
prebiotic potential, 49
preparation, iv, 3, 11, 12, 57, 63
prevalence, 6, 7, 8, 13, 14, 15, 16, 17, 19
prevalence rate, 6
prevention, vii, ix, 7, 22, 24, 25, 26, 27, 28, 30, 32, 34, 37
preventive approach, 32
primary prevention, 22, 28, 32

probiotic, 43, 59, 141, 166
processed meat, 30, 56, 59, 63, 166
protection, 10, 50, 115, 126
proteins, 28, 29, 31, 115
psychiatric disorders, 13
psychology, 96
puberty, 30, 38
public health, 32, 101, 165

Q

qualitative research, 163
quality of life, 13, 28
questionnaire, 6, 16, 19

R

random numbers, 47, 67
rapid socio-ecomonic, 162
recommendations, iv, 9, 26, 159
researchers, 91, 121, 162
resources, 87, 88, 89, 90, 121, 122, 123
risk, vii, viii, ix, 6, 7, 8, 9, 13, 19, 22, 24, 25, 26, 27, 28, 29, 30, 31, 32, 34, 35, 36, 63, 64
risk factors, vii, ix, 22, 26, 27, 36
risk groups, 7, 9, 13

S

safety, 12, 93, 96, 104, 105
saturated fat, ix, 23, 24, 26, 27, 31, 32, 61, 63, 64, 73, 151
saturated fatty acids, 24, 31, 32, 63, 64
saturation, x, 44, 46, 50, 62, 66, 71
sausages, v, vii, ix, 40, 41, 45, 46, 47, 51, 53, 54, 56, 61, 62, 64, 68, 69, 70, 71, 73, 75, 76
school, 23, 78, 83, 97, 98, 105, 154, 155
science, 36, 60, 116, 117, 141

scientific publications, viii, 2
sectoral policies, 136
security, 106, 138, 156
self-discipline, 4
self-efficacy, 33
self-esteem, 4
self-image, 11
sensory acceptance, vii, x, 53, 55, 62, 68
serum, 24, 26, 31, 48
social anxiety, 11
social behavior, 92
social benefits, 13
social change, 98
social problems, 4
social support, 22
society, 10, 13, 96, 100, 162
socioeconomic status, 53
solution, 8, 42, 43, 66, 121, 161
soybean oil, 62, 63, 64, 72, 76
species, 91, 120, 161
spectroscopy, 121
stability, x, 45, 51, 52, 53, 54, 61, 64, 73, 97
storage, 44, 49, 50, 55, 67, 68, 69, 70, 71, 104
stress, 30, 145, 146, 155, 161
sustainability, 109, 110, 111, 129, 136
symptoms, 6, 7, 8, 10, 11, 12, 13, 14
systolic blood pressure, 25

T

techniques, 84, 117, 119, 120, 127, 128, 148, 153
temperature, 43, 45, 62, 65
textural character, 55, 76
texture, vii, ix, x, 39, 41, 47, 50, 51, 53, 54, 55, 61, 62, 67, 68, 71, 72, 73, 76, 117
therapy, 34, 78, 86, 87, 142, 145, 152

traditions, 83, 133, 165
traits, 11
treatment, 8, 19, 24, 26, 44, 45, 65, 92
trial, 23, 33, 36
type 1 diabetes, 34
type 2 diabetes, 34

U

underlying mechanisms, 8
United Kingdom, 91
United States, 91, 98, 102, 113, 130, 147, 156
urban youth, 19

V

vegetable oils, 62, 72, 73, 166
vegetables, 23, 26, 28, 29, 30, 31, 32, 78, 115, 153, 160, 161
vitamins, 31, 87, 88, 89, 90, 161

W

water, ix, x, 40, 41, 42, 44, 45, 46, 48, 50, 51, 53, 55, 56, 62, 65, 66, 68, 75, 84, 114
water absorption, 55
weight control, 8, 10, 88, 89, 90
weight gain, 64
weight loss, 4, 45, 65, 87, 88, 89, 90

Y

yeast, 43, 50, 92
young adults, 38, 103
young people, 6, 97